AF356558

The 40th International Workshop on Bayesian Inference and Maximum Entropy Methods in Science and Engineering

The 40th International Workshop on Bayesian Inference and Maximum Entropy Methods in Science and Engineering

Editors

Wolfgang von der Linden
Sascha Ranftl

MDPI • Basel • Beijing • Wuhan • Barcelona • Belgrade • Manchester • Tokyo • Cluj • Tianjin

Editors
Wolfgang von der Linden
Graz University of Technology
Austria

Sascha Ranftl
Graz University of Technology
Austria

Editorial Office
MDPI
St. Alban-Anlage 66
4052 Basel, Switzerland

This is a reprint of articles from the Conference Proceedings published online in the open access journal *Physical Sciences Forum* (ISSN) (available at: https://www.mdpi.com/2673-9984/3/1).

For citation purposes, cite each article independently as indicated on the article page online and as indicated below:

LastName, A.A.; LastName, B.B.; LastName, C.C. Article Title. *Journal Name* **Year**, *Volume Number*, Page Range.

ISBN 978-3-0365-3200-4 (Hbk)
ISBN 978-3-0365-3201-1 (PDF)

Cover image courtesy of Wolfgang von der Linden.

Contents

About the Editors

Wolfgang von der Linden is a full professor and head of the institute of Theoretical and Computational Physics at TU Graz (Austria). His research focuses on strongly correlated many body physics and Bayesian probability theory, which in 30 years has led to more than 100 papers on this topic. He has collected his wealth of experience together with Volker Dose and Udo von Toussaint, both from the IPP in Garching (Germany) in the book "Bayesian probability theory: applications in the physical sciences", Cambridge University Press, 2014.

Sascha Ranftl is a post-doctoral researcher at the Institute of Theoretical and Computational Physics and Graz Center of Computational Engineering at TU Graz (Austria). His research interests are Bayesian probability theory, physics-informed machine learning, uncertainty quantification for computer simulations and their applications to (biomedical) engineering problems such as the mechanics or detection of aortic dissections.

Preface to "The 40th International Workshop on Bayesian Inference and Maximum Entropy Methods in Science and Engineering"

40-years anniversary: MaxEnt and Bayesian Inference

This workshop series was initiated by Myron Tribus and Edwin T. Jaynes, and organised for the first time by C.Ray Smith and Walter T. Grandy in 1981 at the University in Wyoming. Since then, tremendous progress has been made and the number of publications based on Bayesian methods is impressive.

The workshops are a platform for the presentation of new, sometimes highly revolutionary and ambitious ideas, and for lively and controversial discussions. Some of these ideas have proved very fruitful over the years and have also found their way into other disciplines, while others have proved useless. This year's workshop was also a platform for a mix of ideas that are unanimously considered valuable and correct, and ideas that may have very high potential but are still disputed. We believe that it is in the spirit of the workshop and the conference proceedings to give these ideas a forum for further discussion.

Due to the COVID-19 pandemic, the 2020 workshop had to be cancelled and this year's workshop was held online; however, the lively atmosphere was still unbroken.

In agreement with the general framework of the annual workshop, and due to the broad applicability of Bayesian inference, this year's presentations cover many research areas, such as physics (e.g., plasma physics, astro-physics, statistical mechanics, foundations of quantum mechanics), geodesy, biology, medicine, phonetics, ecology, hydrology, measure theory, image reconstruction, computational engineering, machine learning, and, quite appropriately, epidemiology.

We would like to thank TU Graz and Entropy (MDPI) for their institutional and financial support, as well as Brigitte Schwarz, Stefan Schmutzler and Sabine Pucher for their organizational support. Finally, we thank all authors and speakers as well as the reviewers for their invaluable contributions, and all participants for a constructive and inspiring debate.

Wolfgang von der Linden is a full professor and head of the institute of Theoretical and Computational Physics at TU Graz (Austria). His research focuses on strongly correlated many body physics and Bayesian probability theory, which in 30 years has led to more than 100 papers on this topic. He has collected his wealth of experience together with Volker Dose and Udo von Toussaint, both from the IPP in Garching (Germany) in the book "Bayesian probability theory: applications in the physical sciences", Cambridge University Press, 2014.

Sascha Ranftl is a post-doctoral researcher at the Institute of Theoretical and Computational Physics and Graz Center of Computational Engineering at TU Graz (Austria). His research interests are Bayesian probability theory, physics-informed machine learning, uncertainty quantification for computer simulations and their applications to (biomedical) engineering problems such as the mechanics or detection of aortic dissections.

Wolfgang von der Linden, Sascha Ranftl

Editors

physical sciences forum

Proceeding Paper

Surrogate-Enhanced Parameter Inference for Function-Valued Models [†]

Christopher G. Albert [1,2,*], Ulrich Callies [3] and Udo von Toussaint [1]

[1] Max-Planck-Institut für Plasmaphysik, 85748 Garching, Germany; udt@ipp.mpg.de
[2] Institute of Theoretical and Computational Physics, Technische Universität Graz, 8010 Graz, Austria
[3] Helmholtz-Zentrum Hereon, 21502 Geesthacht, Germany; ulrich.callies@hereon.de
[*] Correspondence: albert@alumni.tugraz.at
[†] Presented at the 40th International Workshop on Bayesian Inference and Maximum Entropy Methods in Science and Engineering, online, 4–9 July 2021.

Abstract: We present an approach to enhance the performance and flexibility of the Bayesian inference of model parameters based on observations of the measured data. Going beyond the usual surrogate-enhanced Monte-Carlo or optimization methods that focus on a scalar loss, we place emphasis on a function-valued output of a formally infinite dimension. For this purpose, the surrogate models are built on a combination of linear dimensionality reduction in an adaptive basis of principal components and Gaussian process regression for the map between reduced feature spaces. Since the decoded surrogate provides the full model output rather than only the loss, it is re-usable for multiple calibration measurements as well as different loss metrics and, consequently, allows for flexible marginalization over such quantities and applications to Bayesian hierarchical models. We evaluate the method's performance based on a case study of a toy model and a simple riverine diatom model for the Elbe river. As input data, this model uses six tunable scalar parameters as well as silica concentrations in the upper reach of the river together with the continuous time-series of temperature, radiation, and river discharge over a specific year. The output consists of continuous time-series data that are calibrated against corresponding measurements from the Geesthacht Weir station at the Elbe river. For this study, only two scalar inputs were considered together with a function-valued output and compared to an existing model calibration using direct simulation runs without a surrogate.

Keywords: parameter inference; Monte Carlo; surrogate model; Gaussian process regression; dimensionality reduction

Citation: Albert, C.G.; Callies, U.; von Toussaint, U. Surrogate-Enhanced Parameter Inference for Function-Valued Models. *Phys. Sci. Forum* **2021**, *3*, 11. https://doi.org/10.3390/psf2021003011

Academic Editors: Wolfgang von der Linden and Sascha Ranftl

Published: 21 December 2021

1. Introduction

Delayed acceptance [1,2] can accelerate Markov chain Monte Carlo (MCMC) sampling up to a factor of one over the acceptance rate. In order to do so, it requires a surrogate of the posterior that contains the cost function inside the likelihood in the case of model calibration. The simplest way to implement delayed acceptance relies on a surrogate with scalar output built for this cost function or for the likelihood. Here, we take an intermediate step and construct a surrogate for the functional output of a blackbox model to be calibrated against reference data. Typical examples are numerical simulations that output time-series or spatial data and depend on tunable input parameters.

There exist numerous related works treating blackbox models with functional outputs with surrogates. Campbell et al. [3] used an adaptive basis of principal component analysis (PCA) to perform global sensitivity analysis. Pratola et al. [4] and Ranjan et al. [5] used GP regression for sequential model calibration in a Bayesian framework. Lebel et al. [6] modeled the likelihood function in an MCMC model calibration via a Gaussian process. Perrin [7] compared the use of a multi-output GP surrogate with a Kronecker structure to an adaptive basis approach.

The present contribution relies on the adaptive basis approach in principal components (Karhunen–Loéve expansion or functional PCA) to reduce the dimensions of the functional output, while modeling the map from inputs to weights in this basis via GP regression. We demonstrate the application of this approach on two examples using usual and hierarchical Bayesian model calibration. In the latter case, a surrogate beyond the L_2 cost function is required if the likelihood depends on additional auxiliary parameters. As an example, we allow variations of the (fractional) order of the norm, thereby, marginalizing over different noise models, including Gaussian and Laplacian noise.

2. Gaussian Process Regression and Bayesian Global Optimization

Gaussian process regression [8–10] is a commonly used tool to construct flexible non-parameteric surrogates. Based on the observed outputs $f(x_k)$ at training points x_k and a covariance function $k(x, x')$, the GP regressor predicts a Gaussian posterior distribution at any point x^*. For a single prediction $f(x^*)$, the expected value and variance of this distribution are given by

$$\bar{f}(x^*) = m(x^*) + K^*(K + \sigma_n I)^{-1} y, \tag{1}$$

$$\mathrm{var}[f(x^*)] = K^{**} - K^*(K + \sigma_n I)^{-1} K^{*T}, \tag{2}$$

where $m(x^*)$ is the mean model, the covariance matrix K contains entries $K_{ij} = k(x_i, x_j)$ based on the training set, $K_i^*(x^*, x_i)$ are entries of a row vector, and $K^{**} = k(x^*, x^*)$ is a scalar. The unit matrix I is added with the noise covariance σ_n, which regularizes the problem and is usually estimated in an optimization loop together with other kernel hyperparameters.

Such a surrogate with uncertainty information can be used for Bayesian global optimization [11–13] of the log-posterior as a cost function. Here, we apply this method to reach the vicinity of the posterior's mode before sampling. As an acquisition function, we use the expected improvement (see, e.g., [12]) at a newly observed location x^* given existing training data $\mathcal{D}$,

$$a_{\mathrm{EI}}(x^\star) = E[\max(0, \bar{f}(x^*) - \hat{f}) | x^*, \mathcal{D}]$$
$$= (\bar{f}(x^*) - \hat{f}) \Phi(\hat{f}; \bar{f}(x^*), \mathrm{var}[f(x^*)]) + \mathrm{var}[f(x^*)] \mathcal{N}(\hat{f}; \bar{f}(x^*), \mathrm{var}[f(x^*)]), \tag{3}$$

where $\hat{f}$ is the optimum value for $f(x)$ observed thus far. Due to the non-linear transformation from the functional blackbox output to the value of the cost function, it is more convenient to realize Bayesian optimization with a direct GP surrogate of the cost function that is constructed in addition to the surrogate for the functional output for the KL expansion coefficients described below.

3. Delayed Acceptance MCMC

Delayed acceptance MCMC builds on a fast surrogate for the posterior $\tilde{p}(x|y)$ to reject unlikely proposals early [1,2]. Following the usual Metropolis–Hastings algorithm, the probability to accept a new proposal x^* in this first stage in the n-the step of the Markov chain is, as usual,

$$\tilde{P}_{\mathrm{acc}}^n = \frac{\tilde{p}(x^*|y)}{\tilde{p}(x_{n-1}|y)} \frac{g(x_{n-1}|x^*)}{g(x^*|x_{n-1})}, \tag{4}$$

where g is a transition probability that has been suitably tuned during warmup. The true posterior $p(x|y)$ is only evaluated if the proposal "survives" this first stage and enters the final acceptance probability

$$P_{\mathrm{acc}}^n = \frac{p(x^*|y)}{p(x_{n-1}|y)} \frac{\tilde{p}(x_{n-1}|y)}{\tilde{p}(x^*|y)}. \tag{5}$$

Actual computation is typically performed in the logarithmic space with a cost function

$$\ell(x|y) \equiv -\log p(x|y). \tag{6}$$

If this function is fixed, it is most convenient to directly build a surrogate $\tilde{\ell}(x|y)$ for the log-posterior $\ell(x|y)$ including the corresponding prior.

4. Bayesian Hierarchical Models and Fractional Norms

One application of modeling the full functional output instead of only the cost function is the existence of additional distribution parameters θ in the likelihood in addition to the original model inputs x. Such dependencies appear within Bayesian hierarchical models [14], where θ are again subject to a certain (prior) distribution with possibly further levels of hyperparameters. There are essentially two ways to construct a surrogate with support for additional parameters θ: Building a surrogate for the cost function that adds θ as independent variables or constructing a surrogate with functional output for $f_k(x)$ and keeping the dependencies on θ exact. Here, we focus on the latter, and apply this surrogate within delayed acceptance MCMC with both, x and θ as tunable parameters.

As an example, we use a more general noise model than the usual Gaussian likelihood that builds on arbitrary ℓ^θ norms [15–17] with real-valued θ not fixed while traversing the Markov chain. We allow members of the exponential family for observational noise and specify only its scale, but keep θ as a free parameter. Namely, we model the likelihood for observing y in the output as

$$p(y|x,\theta) = \frac{1}{2\sqrt{2}\sigma\,\Gamma(1+\theta^{-1})}e^{-\ell(y;x,\theta)}, \tag{7}$$

with the normalized ℓ^θ norm to the power of θ,

$$\ell(y;x,\theta) \equiv \frac{1}{D}\sum_{i=1}^{D}\left|\frac{y_i - f_i(x)}{\sqrt{2}\sigma}\right|^\theta \tag{8}$$

as the loss function between observed data y_i and blackbox model $f_i(x)$. Choosing the usual L_2 norm leads to a Gaussian likelihood for the noise model, whereas using the L_1 norm means Laplacian noise. To maintain the relative scale when varying θ, it is important to add the term $\log\Gamma(1+\theta^{-1})$ from (7) to the negative log-likelihood. In the following use cases, we are going to compare the cases of fixed and variable θ.

5. Linear Dimension Reduction via Principal Components

Formally, the blackbox output for given input x can be a function $f(t) \in \mathbb{H}$ in an infinite-dimensional Hilbert space (though sampled at a finite number of points in practice). Linear dimension reduction in such a space means finding the optimum set of basis functions $\varphi_k(t)$ that spans the output space $f(t;x)$ for any input x given to the blackbox. The reduced model of order r is then given by

$$f(t;x) \approx \sum_{k=1}^{r} z_k(x)\varphi_k(t). \tag{9}$$

This approach is known as the Karhunen–Loéve (KL) expansion [18] in case $f(t;x)$ are interpreted as realizations of a random process, or as the functional principal component analysis (FPCA) [19]. For our application, this distinction does not matter. The KL expansion boils down to solving a regression problem in the non-orthogonal basis of N observed realizations to represent new observations. Then, an eigenvalue problem is solved to invert the $N \times N$ collocation matrix with entries

$$M_{ij} = \langle f(t;x_i), f(t;x_j)\rangle. \tag{10}$$

Here, the inner product in Hilbert spaces and its approximation for a finite set of support points is given by

$$\langle u, v \rangle = \int_\Omega u(t)v(t)\, \mathrm{d}t \approx \frac{1}{N_t} \sum_{k=1}^{N_t} u(t_k)v(t_k). \tag{11}$$

If $N_t \gg N$ (many support points, few samples), solving the eigenvalue problem of the collocation matrix M is more efficient than the dual one of the covariance matrix C with $C_{ij} = \sum_k f(t_i, x_k)f(t_j, x_k)$ in the usual PCA (see [9] for their equivalence via the singular value decomposition of $Y_{ij} = f(t_i, x_j)$). The question of at which r to truncate the eigenspectrum in (9) depends on the desired accuracy in the output, which is briefly analyzed in the following paragraph.

Error Estimate

Here, we justify why we can assume an L_2 truncation error of the order of the ratio λ_r / λ_1 between the smallest eigenvalue considered in the approximation and the largest one. The truncated SVD can be shown to be the best linear approximation $M^{(r)}$ of lower rank r to an $N \times N$ matrix M in terms of the Frobenius norm $||M||_\mathrm{F}$ (see, e.g., [20]). Its value is simply computed from the L_2 norm of singular values,

$$||M||_\mathrm{F} = \left(\sum_{k=1}^{N} \sigma_k^2 \right)^{1/2}, \tag{12}$$

where $\sigma_k^2 = \lambda_k$ in the case of real eigenvalues λ_k of a positive semi-definite matrix as for the covariance or collocation matrix. The truncation error is given by

$$||M^{(r)} - M||_\mathrm{F} = \left(\sum_{k=r+1}^{N} \lambda_k \right)^{1/2}. \tag{13}$$

The error estimate for the KL expansion uses this convenient property together with the fact that the Frobenius norm is compatible with the usual L_2 norm $|x|$ of vectors y, i.e.,

$$|My| \le ||M||_\mathrm{F}|y|. \tag{14}$$

Representing y via the first r eigenvalues of the collocation matrix yields a relative squared reconstruction error of

$$|(M^{(r)} - M)y|^2 / |y|^2 \le \sum_{k=r+1}^{N} \lambda_k \le (N - r)\lambda_r. \tag{15}$$

The last estimate is relatively crude if $N \gg r$, and the spectrum decays fast with the index variable k. If one assumes a decay rate α with

$$\lambda_k \approx \lambda_r(k - r)^{-\alpha}, \tag{16}$$

one obtains

$$\sum_{k=r+1}^{N} \lambda_k \approx \sum_{k=r+1}^{\infty} \lambda_r(k - r)^{-\alpha} = \lambda_r \sum_{k=1}^{\infty} k^{-\alpha} = \lambda_r \zeta(\alpha), \tag{17}$$

where ζ is the Riemann zeta function. This function diverges for a spectral decay of order $\alpha = 1$ and reaches its asymptotic value $\zeta(\infty) = 1$ relatively quickly for $\alpha \ge 2$ (e.g., $\zeta(3) = 1.2$). The spectral decay rate α can be fitted in a log–log plot of λ_k over index k and takes values between $\alpha = 3$ and 5 in our use case. The underlying assumptions are violated if the spectrum stagnates at a large number of constant eigenvalues for higher indices k.

6. Implementation and Results

The idea behind the realization of MCMC with a function-valued surrogate is quite simple. Instead of directly using the surrogate for the cost ℓ with fixed θ, we take a step in-between. Multiple surrogates $\tilde{z}_k(x)$ are built, where each maps the input x to one weight $z_k(x)$ in the KL expansion. A surrogate $\tilde{f}_i(x) \equiv \tilde{f}(t_i; x)$ for the model output is then given by replacing $z_k(x)$ by $\tilde{z}_k(x)$ in (9). The according surrogate $\tilde{\ell}(y; x, \theta)$ for the cost function uses $\tilde{f}_i(x)$ instead of $f_i(x)$ in (8). Dependencies on θ are kept exact in this approach. The main algorithm proceeds in the following steps:

1. Construct a GP surrogate for the L_2 cost function on a space-filling sample sequence over the whole prior range.
2. Refine the sampling points near the posterior's mode through Bayesian global optimization with the L_2 cost surrogate.
3. Train a multi-output GP surrogate for the functional output $z(x)$ on the refined sampling points.
4. Use the function-valued surrogate for delayed acceptance in the MCMC run.

For all GP surrogates, we use a Matern 5/2 kernel for $k(x, x')$ together with a linear mean model for $m(x)$, as realized in the Python package GPy [21]. For step 4, we use Gibbs sampling and the surrogate for $z(x)$, yielding the full output $y(t, x)$ rather than only the L_2 distance to a certain reference dataset. The idea to refine the surrogate iteratively during MCMC had to be abandoned early. The problem is that detailed balance is violated as soon as the surrogate proposal probabilities change when modifying the GP regressor with a new point. In the following application cases, we compare a usual MCMC evaluation using the full model to MCMC with delayed acceptance using the GP surrogate together with the KL expansion/functional PCA (GP+KL) in the output function space.

6.1. Toy Model

First, we test the quality of the algorithm on a toy model given by

$$y(t, x) = x_1 \sin((t - x_2)^3). \tag{18}$$

We choose reference values $x_1 = 1.15$, $x_2 = 1.4$ to test the calibration of x against the according output $y^{\text{ref}}(t) \equiv y(t, x^{\text{ref}})$ and add Gaussian noise of amplitude $\sigma = 0.05$. A flat prior is used for x. For the hierarchical model case (7), we choose a starting guess of $\theta = 2$ for the norm's order and a Gaussian prior with $\sigma_\theta = 0.5$ around this value together with a positivity constraint. The initial sampling domain in the square $x_1, x_2 \in (0, 2)$. The comparison between MCMC and delayed acceptance MCMC is made once for fixed $\theta = 2$ (Gaussian likelihood) and then for a hierarchical model with a random walk also in θ. The respective Markov chain with 10,000 steps has a correlation length of ≈ 10 steps (Figure 1) and yields a posterior parameter distribution for (x_1, x_2) depicted in Figure 2.

The results in Figure 2 show good agreement in the posterior distributions of full MCMC and delayed acceptance MCMC. Compared to the case with fixed $\theta = 2$, the additional freedom in θ in the hierarchical model leads to further exploration of the parameter space. The posterior of θ according to the Markov chain is given in Figure 3. The similarity to the prior distribution shows that the data does not yield new information on how to choose θ.

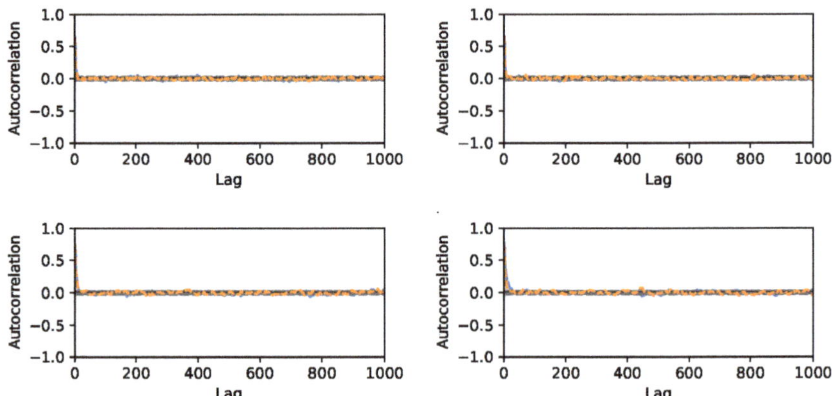

Figure 1. Autocorrelation over lag in MCMC steps for inputs x_1 (solid) and x_2 (dashed) in the toy model. **Top**: Gaussian likelihood, and **bottom**: hierarchical model. **Left**: full MCMC, and **right**: delayed acceptance MCMC with GP+KL surrogate.

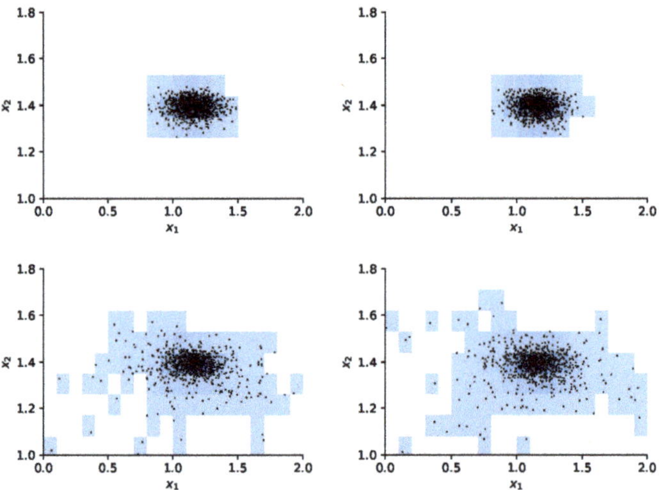

Figure 2. Posterior distribution of the calibrated parameters x in (18). **Top**: Gaussian likelihood, **bottom**: hierarchical model. **Left**: full MCMC, **right**: delayed acceptance MCMC with GP+KL surrogate.

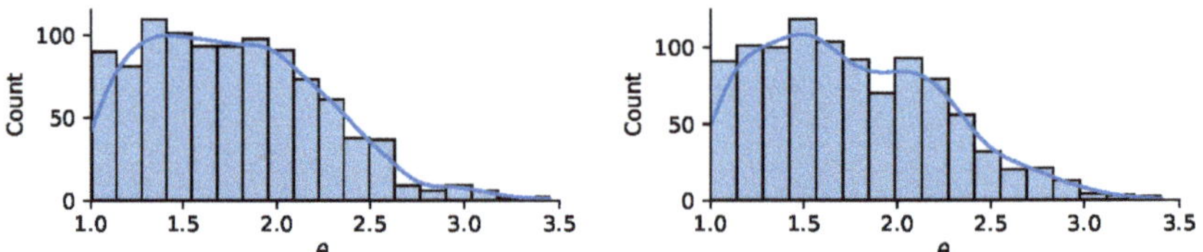

Figure 3. Posterior distribution of the fractional order θ in the loss function with ℓ^θ norm. **Left**: full MCMC, **right**: delayed acceptance MCMC with GP+KL surrogate.

6.2. Riverine Diatom Model

The final application of the described method is on a riverine diatom model [22,23]. This model predicts the chlorophyll a concentration at an observation point at the Elbe river as a time series and depends on several input parameters. For simplicity, and to limit computational resources, we select only two of the six scalar inputs and use fixed values for the remaining four. Namely, the chosen parameters $x_1 = K_{\text{light}}$ and $x_2 = \mu_0$ appear in the growth rate inside the diatom model. The latter is given by the "Smith formula" [24] for photosynthesis,

$$\mu(t) \propto \mu_0 \frac{1}{D} \int_0^D \frac{I(t)e^{-\lambda(t)z}}{\sqrt{K_{\text{light}}^2 + I^2(t)e^{-2\lambda(t)z}}} dz,$$

where D is the water depth, and $I(t)$ is the radiation intensity prescribed at the water surface. Light attenuation $\lambda(t) \equiv \lambda_S C_{\text{chl}}(t)$ is modeled to be proportional to the chlorophyll a concentration $C_{\text{chl}}(t)$. Equations are solved within a Lagrangian setup, following water parcels that travel down the Elbe river. Data points of the local chlorophyll time series simulated at Geesthacht Weir are made up by chlorophyll a values at the Lagrangian trajectory end points. These values are the functional model output $y(t)$ for which the model is calibrated with respect to measurements $y_{\text{ref}}(t)$. As the parameters are positive and limited by reasonable maximum values from domain knowledge, we use a half-sided Cauchy (Lorentz) prior

$$p(x_k) = \frac{2}{\pi} \frac{b_k}{b_k^2 + x_k^2} \text{ for } x_k > 0, \quad p(x_k) = 0 \text{ for } x_k \le 0. \tag{19}$$

Here, we choose a scale value x_k^* for which $P^* = 90\%$ of the probability volume is contained within $x_k < x_k^*$. Considering the cumulative distribution, we have to set

$$b_k = \frac{x_k^*}{\tan\left(\frac{\pi}{2} P^\star\right)} \tag{20}$$

to realize this condition.

As in the case of the toy model, we use 10,000 steps in the Markov chain. The results for autocorrelation and posterior samples using the full model versus delayed acceptance are shown in Figures 4 and 5. The correlation time of $\approx$500 steps is much larger than in the toy model, and the decay of the autocorrelation over the lag roughly matches between the two approaches. Delayed acceptance sampling produces similar posterior samples in Figure 5 at about one third of the overall computation time. There, one also sees the issue of high correlation between K_{light} and μ_0 in the posterior of the calibration, making Gibbs sampling inefficient in that particular case.

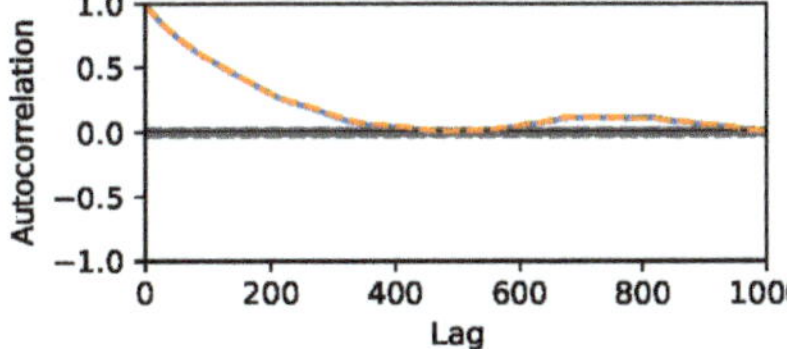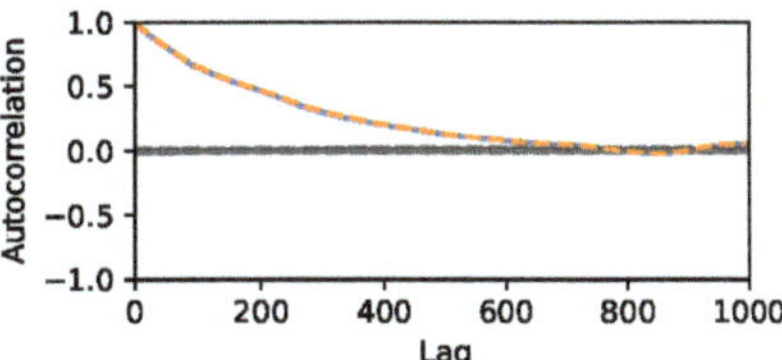

Figure 4. Autocorrelation over lag in MCMC steps for inputs K_{light} (solid) and μ_0 (dashed) in the riverine diatom model. **Top:** Gaussian likelihood, **bottom:** hierarchical model. **Left:** full MCMC, **right:** delayed acceptance MCMC with GP + KL surrogate.

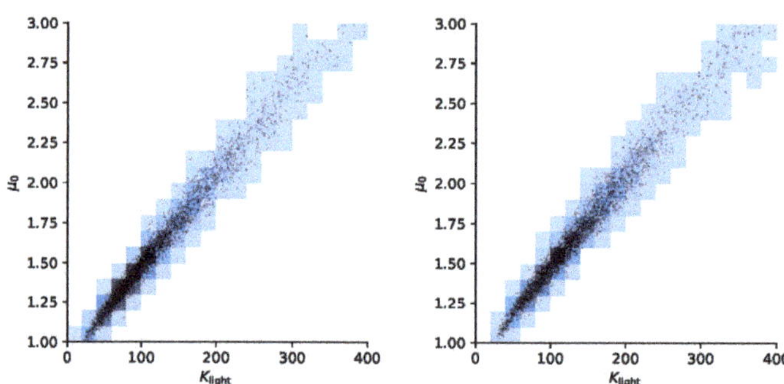

Figure 5. Posterior distribution of calibrated parameters for the riverine diatom model. **Left:** full MCMC, **right:** delayed acceptance MCMC with GP + KL surrogate.

7. Conclusions and Outlook

We illustrated the application of function-valued surrogates to delayed acceptance MCMC for parameter calibration in simple as well as hierarchical Bayesian models. Using a surrogate for the functional output rather than a cost function or likelihood is useful for several reasons. Conceptually, it allows introducing additional distribution parameters in Bayesian hierarchical models. Our results demonstrate that it is possible and efficient to perform MCMC with delayed acceptance on such models while keeping the dependencies in these additional parameters exact. In particular, the fractional order of the norm appearing in the cost function was left free, which is useful for robust model calibration.

The method was applied to a toy model and an application case of a riverine diatom model. In both cases, using delayed acceptance with a surrogate for the functional output produced results comparable to using the full model at only about one third of the actual model evaluations. Compared to direct surrogate modeling of the cost function, we could also observe an increase in the quality of the predicted cost. This is likely connected to the higher flexibility of modeling weights to multiple principal components with Gaussian processes with individual hyperparameters.

The described approach is not immune to the curse of dimensionality. On the one hand, the number of required GP regressors grows linearly with the effective dimensions of the *output* function space. Since evaluation is fast and parallelizable, this is a minor issue in practice. On the other hand, increasing the dimension of the *input* space soon prohibits the construction of a reliable surrogate due to the required number training points to fill the parameter space. In such cases, the preprocessing overhead is expected to outweigh the speedup of delayed acceptance MCMC for either functional or scalar surrogates. More detailed investigations will be required to give quantitative estimates on this trade off.

Institutional Review Board Statement: Not applicable.

Informed Consent Statement: Not applicable.

Data Availability Statement: Data available in a publicly accessible repository. The data presented in this study are openly available in Zenodo at https://doi.org/10.5281/zenodo.5773865, reference number [25].

Acknowledgments: This study is a contribution to the *Reduced Complexity Models* grant number ZT-I-0010 funded by the Helmholtz Association of German Research Centers.

Conflicts of Interest: The authors declare no conflict of interest.

References

1. Christen, J.A.; Fox, C. Markov Chain Monte Carlo Using an Approximation. *J. Comput. Graph. Stat.* **2005**, *14*, 795–810. [CrossRef]
2. Wiqvist, S.; Picchini, U.; Forman, J.L.; Lindorff-Larsen, K.; Boomsma, W. Accelerating Delayed-Acceptance Markov Chain Monte Carlo Algorithms. *arXiv* **2019**, arXiv:1806.05982.
3. Campbell, K.; McKay, M.D.; Williams, B.J. Sensitivity Analysis When Model Outputs Are Functions. *Reliab. Eng. Syst. Saf.* **2006**, *91*, 1468–1472. [CrossRef]
4. Pratola, M.T.; Sain, S.R.; Bingham, D.; Wiltberger, M.; Rigler, E.J. Fast Sequential Computer Model Calibration of Large Nonstationary Spatial-Temporal Processes. *Technometrics* **2013**, *55*, 232–242. [CrossRef]
5. Ranjan, P.; Thomas, M.; Teismann, H.; Mukhoti, S. Inverse Problem for a Time-Series Valued Computer Simulator via Scalarization. *Open J. Stat.* **2016**, *6*, 528–544. [CrossRef]
6. Lebel, D.; Soize, C.; Fünfschilling, C.; Perrin, G. Statistical Inverse Identification for Nonlinear Train Dynamics Using a Surrogate Model in a Bayesian Framework. *J. Sound Vib.* **2019**, *458*, 158–176. [CrossRef]
7. Perrin, G. Adaptive Calibration of a Computer Code with Time-Series Output. *Reliab. Eng. Syst. Saf.* **2020**, *196*, 106728. [CrossRef]
8. O'Hagan, A. Curve Fitting and Optimal Design for Prediction. *J. R. Stat. Soc. Ser. B* **1978**, *40*, 1–24. [CrossRef]
9. Bishop, C.M. *Pattern Recognition and Machine Learning*; Springer: Berlin, Germany, 2006.
10. Rasmussen, C.E.; Williams, C.K.I. *Gaussian Processes for Machine Learning*; MIT Press: Cambridge, MA, USA, 2006. [CrossRef]
11. Shahriari, B.; Swersky, K.; Wang, Z.; Adams, R.P.; de Freitas, N. Taking the Human Out of the Loop: A Review of Bayesian Optimization. *Proc. IEEE* **2016**, *104*, 148–175. [CrossRef]
12. Osborne, M.A.; Garnett, R.; Roberts, S.J. *Gaussian Processes for Global Optimization*; Learning and 3rd International Conference on Learning and Intelligent Optimization, LION 3; Springer: Trento, Italy, 2009.
13. Preuss, R.; von Toussaint, U. Global Optimization Employing Gaussian Process-Based Bayesian Surrogates. *Entropy* **2018**, *20*, 201. [CrossRef] [PubMed]
14. Allenby, G.M.; Rossi, P.E.; McCulloch, R.E. *Hierarchical Bayes Models: A Practitioners Guide*; SSRN Scholarly Paper ID 655541; Social Science Research Network: Rochester, NY, USA, 2005. [CrossRef]
15. Aggarwal, C.C.; Hinneburg, A.; Keim, D.A. *On the Surprising Behavior of Distance Metrics in High Dimensional Space*; Database Theory—ICDT 2001; Lecture Notes in Computer Science; Van den Bussche, J.; Vianu, V., Eds.; Springer: Berlin/Heidelberg, Germany, 2001; pp. 420–434. [CrossRef]
16. Dose, V. Bayesian Estimate of the Newtonian Constant of Gravitation. *Meas. Sci. Technol.* **2006**, *18*, 176–182. [CrossRef]
17. Flexer, A.; Schnitzer, D. Choosing Lp Norms in High-Dimensional Spaces Based on Hub Analysis. *Neurocomputing* **2015**, *169*, 281–287. [CrossRef]
18. Newman, A.J. *Model Reduction via the Karhunen-Loeve Expansion Part I: An Exposition*; Institute for Systems Research Technical Reports; Univ. Maryland: College Park, MD, USA, 1996.
19. Shang, H.L. A Survey of Functional Principal Component Analysis. *AStA Adv. Stat. Anal.* **2014**, *98*, 121–142. [CrossRef]
20. Cadzow, J.A. Spectral Analysis. In *Handbook of Digital Signal Processing*; Elsevier: Amsterdam, The Netherlands, 1987; pp. 701–740.
21. GPy. *GPy: A Gaussian Process Framework in Python*; Software Publication, Univ. Sheffield: Sheffield, UK, 2012.
22. Callies, U.; Scharfe, M.; Ratto, M. Calibration and Uncertainty Analysis of a Simple Model of Silica-Limited Diatom Growth in the Elbe River. *Ecol. Model.* **2008**, *213*, 229–244. [CrossRef]
23. Scharfe, M.; Callies, U.; Blöcker, G.; Petersen, W.; Schroeder, F. A Simple Lagrangian Model to Simulate Temporal Variability of Algae in the Elbe River. *Ecol. Model.* **2009**, *220*, 2173–2186. [CrossRef]
24. Smith, E.L. Photosynthesis in Relation to Light and Carbon Dioxide. *Proc. Natl. Acad. Sci. USA* **1936**, *22*, 504–511. [CrossRef] [PubMed]
25. Albert, C.G.; von Toussaint, U.; Callies, U. Dataset for article "Surrogate-Enhanced Parameter Inference for Function-Valued Models", v1.0. Zenodo, 2021.

physical sciences forum

Proceeding Paper

Quantum Mechanics as Hamilton–Killing Flows on a Statistical Manifold [†]

Ariel Caticha

Physics Department, University at Albany-SUNY, Albany, NY 12222, USA; acaticha@albany.edu

† Presented at the 40th International Workshop on Bayesian Inference and Maximum Entropy Methods in Science and Engineering, online, 4–9 July 2021.

Abstract: The mathematical formalism of quantum mechanics is derived or "reconstructed" from more basic considerations of the probability theory and information geometry. The starting point is the recognition that probabilities are central to QM; the formalism of QM is derived as a particular kind of flow on a finite dimensional statistical manifold—a simplex. The cotangent bundle associated to the simplex has a natural symplectic structure and it inherits its own natural metric structure from the information geometry of the underlying simplex. We seek flows that preserve (in the sense of vanishing Lie derivatives) both the symplectic structure (a Hamilton flow) and the metric structure (a Killing flow). The result is a formalism in which the Fubini–Study metric, the linearity of the Schrödinger equation, the emergence of complex numbers, Hilbert spaces and the Born rule are derived rather than postulated.

Keywords: information geometry; symplectic geometry; Hamilton–Killing flows; entropic dynamics

Citation: Caticha, A. Quantum Mechanics as Hamilton–Killing Flows on a Statistical Manifold. *Phys. Sci. Forum* **2021**, *3*, 12. https://doi.org/10.3390/psf2021003012

Academic Editors: Wolfgang von der Linden and Sascha Ranftl

Published: 21 December 2021

Publisher's Note: MDPI stays neutral with regard to jurisdictional claims in published maps and institutional affiliations.

1. Introduction

In the traditional approach to quantum mechanics (QM), the Hilbert space plays a central, dominant role and probabilities are introduced, almost as an afterthought, in order to provide the phenomenological link for handling measurements. The uneasy coexistence of the Hilbert and the probabilistic structures is reflected in the two separate modes of wave-function evolution; one is the linear and deterministic Schrödinger evolution and the other is the discontinuous and stochastic wave function collapse. It has given rise to longstanding problems in the interpretation of the quantum state itself [1–5].

These difficulties have motivated alternative approaches in which, rather than postulating Hilbert spaces as the starting point, one recognizes that probabilities play the dominant role; probabilities are not just an accidental feature peculiar to quantum measurements. The goal there is to derive or "reconstruct" the mathematical formalism of QM from more basic considerations of probability theory and geometry. (See, e.g., [6–11] and references therein.)

In the entropic dynamics (ED) approach the central object is the epistemic configuration space, which is a statistical manifold—a space in which each point represents a probability distribution [11]. In this paper, our goal is to discuss those special curves that could potentially play the role of trajectories. What makes those curves special is that they are adapted to the natural geometric structures on the statistical manifold.

Two such structures are of central importance. The first is familiar from statistics, i.e., all statistical manifolds have an intrinsic metric structure given by the information metric [12,13]. The second is familiar from classical mechanics [14–16]. Since we are interested in trajectories, we are naturally led to consider the vectors that are tangent to such curves, as well as the dual vectors, or covectors—it is these objects that are used to represent the analogues of the velocities of probabilities and their momenta. Vectors and covectors live in the so-called tangent and cotangent spaces, respectively. It turns out that the statistical manifold plus all its cotangent spaces is itself a manifold—the cotangent bundle—that can be endowed with a second natural structure called *symplectic*. In mechanics, the cotangent bundle is known as phase space and the symplectic transformations are known as canonical transformations.

There is extensive literature on the symplectic and metric structures inherent to QM. They have been discovered, independently rediscovered, and extensively studied by many authors [17–26]. Their crucial insight is that those structures, being of purely geometrical nature, are not just central to classical mechanics, they are also central to quantum mechanics. Furthermore, the potential connection and relevance of information geometry to various aspects of QM, including its metric structure, has also been studied [9–11,27–30].

To characterize congruences of curves in the epistemic phase space—or, equivalently, the *flows* on the cotangent bundle—we must address two problems. First, we must characterize the particular cotangent space and the symplectic structure that is relevant to QM. This amounts to establishing the correct conjugate momenta to be paired to the coordinates, which, in our case, are probabilities. In classical mechanics, this pairing is accomplished with the help of a Lagrangian $L(q, \dot{q})$ and the prescription $p = \partial L / \partial \dot{q}$. In the present problem, we have no access to a Lagrangian and a different criterion is adopted [11]. The second problem is to provide the cotangent bundle with a metric structure that is compatible with the information metric of the underlying statistical manifold. The issue is that cotangent bundles are not statistical manifolds and the challenge is to identify the natural set of assumptions that leads to the right metric structure.

We show that the flows that are relevant to quantum mechanics are those that preserve (in the sense of vanishing Lie derivatives) both the symplectic structure (a Hamilton flow) and the metric structure (a Killing flow). The characterization of these Hamilton–Killing (HK) flows results in a formalism that includes states described by rays, a geometry given by the Fubini–Study metric, flows that obey a linear Schrödinger equation, the emergence of a complex structure, the Born rule, and Hilbert spaces. All these elements are derived rather than postulated.

The present discussion includes two new developments. First, our focus is on isolating the essential geometrical aspects of the problem (a discussion of the physical aspects is given in [11]) and the main ideas are presented in the simpler context of a finite-dimensional manifold—a simplex. Thus, what we derive here is the geometrical framework that applies to a toy model—an n-sided quantum die. Second, the metric structure of the cotangent bundle is found by a new argument involving the minimal assumption that the metric of phase space is determined by the only metric structure at our disposal, namely, the information metric of the simplex.

Is this all there is to quantum mechanics? We conclude with a word of caution. The framework developed here takes us a long way towards justifying the mathematical formalism that underlies quantum mechanics, but it is only a kinematical prelude to the true dynamics. The point is that not every HK curve is a trajectory and not every parameter that labels points along a curve is time. All changes of probabilities, including the changes we call dynamics, must be compatible with the entropic and Bayesian rules that have been found to be of universal applicability in inference. It is this additional requirement that further restricts the HK flows to an entropic dynamics that describes an evolution in a suitably constructed entropic concept of time [7,11].

This paper focuses on deriving the mathematical formalism of quantum mechanics, but the ED approach has been applied to a variety of other topics in quantum theory. These include the quantum measurement problem [31,32]; momentum and uncertainty relations [33,34]; the Bohmian limit [34,35] and the classical limit [36]; extensions to curved spaces [37]; to relativistic fields [38–40]; and the ED of spin [41].

2. Some Background

We deal with several distinct spaces. One is the *ontic configuration space* of microstates labeled by $i = 1 \ldots n$, which are the unknown variables we are trying to predict. Another is the space of probability distributions $\rho = (\rho^1 \dots \rho^n)$, which is the *epistemic configuration space* or, to use a shorter name, the *e-configuration space*. This $(n-1)$-dimensional statistical manifold is a simplex $\mathcal{S}$,

$$\mathcal{S} = \left\{ \rho \mid \rho^i \geq 0; \ \textstyle\sum_{i=1}^{n} \rho^i = 1 \right\}. \tag{1}$$

As coordinates for a generic point ρ on $\mathcal{S}$, we shall use the probabilities ρ^i themselves.

Given the manifold $\mathcal{S}$, we can construct two other special manifolds that will turn out to be useful, the *tangent bundle* $T\mathcal{S}$ and the *cotangent bundle* $T^*\mathcal{S}$. These are fiber bundles; the *base manifold* is $\mathcal{S}$ and the fibers at each point ρ are respectively the tangent $T\mathcal{S}_\rho$ and cotangent $T^*\mathcal{S}_\rho$ spaces at ρ. The tangent space at ρ, $T\mathcal{S}_\rho$, is the vector space composed of all vectors that are tangent to curves through the point ρ. While this space is obviously important (it is the space of "velocities" of probabilities), in what follows, we will not have much to say about it. Much more central to our discussion is the cotangent space at ρ, $T^*\mathcal{S}_\rho$ which is the vector space of all covectors at ρ.

As already mentioned, the reason we care about vectors and covectors is that these are the objects that are used to represent velocities and momenta. The cotangent bundle $T^*\mathcal{S}$, plays the central role of the *epistemic phase space*, or *e-phase space*.

A point $X \in T^*\mathcal{S}$ is represented as $X = (\rho, \pi)$, where $\rho = (\rho^1 \dots \rho^n)$ are coordinates on the base manifold $\mathcal{S}$ and $\pi = (\pi_1 \dots \pi_n)$ are some generic coordinates on the cotangent space $T^*\mathcal{S}_\rho$ at ρ. Curves on $T^*\mathcal{S}$ allow us to define vectors on the tangent spaces $T(T^*\mathcal{S})_X$. Let $X = X(\lambda)$ be a curve parameterized by λ; then, the vector $\bar{V}$ tangent to the curve at $X = (\rho, \pi)$ has components $d\rho^i / d\lambda$ and $d\pi_i / d\lambda$ and is written as

$$\bar{V} = \frac{d}{d\lambda} = \frac{d\rho^i}{d\lambda} \bar{\rho}_i + \frac{d\pi_i}{d\lambda} \bar{\pi}^i = \frac{d\rho^i}{d\lambda} \frac{\partial}{\partial \rho^i} + \frac{d\pi_i}{d\lambda} \frac{\partial}{\partial \pi_i}, \tag{2}$$

where $\bar{\rho}_i$ and $\bar{\pi}^i$ are the basis vectors, the index $i = 1 \dots n$ is summed over and we adopt the standard notation in differential geometry, $\bar{\rho}_i = \partial/\partial \rho^i$ and $\bar{\pi}^i = \partial/\partial \pi_i$. The directional derivative of a function $F(X)$ along the curve $X(\lambda)$ is

$$\frac{dF}{d\lambda} = \frac{\partial F}{\partial \rho^i} \frac{d\rho^i}{d\lambda} + \frac{\partial F}{\partial \pi_i} \frac{d\pi_i}{d\lambda} \stackrel{\text{def}}{=} \tilde{\nabla} F[\bar{V}], \tag{3}$$

where $\tilde{\nabla}$ is the gradient in $T^*\mathcal{S}$, that is, the gradient of a generic function $F(X) = F(\rho, \pi)$ is

$$\tilde{\nabla} F = \frac{\partial F}{\partial \rho^i} \tilde{\nabla} \rho^i + \frac{\partial F}{\partial \pi_i} \tilde{\nabla} \pi_i, \tag{4}$$

where $\tilde{\nabla} \rho^i$ and $\tilde{\nabla} \pi_i$ are the basis covectors and the tilde '~' serves to distinguish the gradient $\tilde{\nabla}$ on the bundle $T^*\mathcal{S}$ from the gradient ∇ on the simplex $\mathcal{S}$.

Here, unfortunately, we encounter a technical difficulty due to the fact that the space $\mathcal{S}$ is constrained to normalized probabilities so that the coordinates ρ^i cannot be varied independently. This problem is handled, without loss of generality, by embedding the $(n-1)$-dimensional manifold $\mathcal{S}$ into a manifold of one dimension higher, the so-called positive-cone, denoted $\mathcal{S}^+$, where the coordinates ρ^i are unconstrained.

To simplify the notation, a point $X = (\rho, \pi)$ in the $2n$-dimensional $T^*\mathcal{S}^+$ is labeled by its coordinates $X^{\alpha i} = (X^{1i}, X^{2i}) = (\rho^i, \pi_i)$, where αi is a composite index. The first index α (chosen from the beginning of the Greek alphabet) takes two values, $\alpha = 1, 2$. Since α keeps track of whether i is an upper ρ^i index ($\alpha = 1$) or a lower π_i index ($\alpha = 2$), from now on we can set $\rho_i = \rho^i$. Then, Equations (2) and (4) are written as

$$\bar{V} = \frac{d}{d\lambda} = V^{\alpha i} \frac{\partial}{\partial X^{\alpha i}} \quad \text{and} \quad \tilde{\nabla} F = \frac{\partial F}{\partial X^{\alpha i}} \tilde{\nabla} X^{\alpha i}. \tag{5}$$

The repeated indices indicate a double summation over α and i. The action of the basis covectors $\tilde{\nabla} X^{\alpha i}$ on the basis vectors, $\partial/\partial X^{\beta j} = \partial_{\beta j}$, is given by

$$\tilde{\nabla} X^{\alpha i}[\partial_{\beta j}] = \frac{\partial X^{\alpha i}}{\partial X^{\beta j}} = \delta^{\alpha i}_{\beta j} \quad \text{so that} \quad \tilde{\nabla} F[\bar{V}] = \frac{\partial F}{\partial X^{\alpha i}} V^{\alpha i} = \frac{dF}{d\lambda} \tag{6}$$

is the directional derivative of F along the vector $\bar{V}$.

3. Hamiltonian Flows

Just as a manifold can be supplied with a symmetric bilinear form, the metric tensor, which gives it the fairly rigid structure described as its *metric geometry*, cotangent bundles can be supplied with an *antisymmetric* bilinear form, the *symplectic form*, which gives them the somewhat floppier structure called *symplectic geometry* (Arnold 1997 [15,16]).

A vector field $\bar{V}(X)$ defines a space-filling congruence of curves $X^i = X^i(\lambda)$ that are tangent to the field $\bar{V}(X)$ at every point X. We seek those special congruences or *flows* that reflect the symplectic geometry.

3.1. hE Symplectic Form

Once the local coordinates (ρ^i, π_i) on $T^*\mathcal{S}^+$ are established there is a natural choice of symplectic form

$$\Omega = \tilde{\nabla}\rho^i \otimes \tilde{\nabla}\pi_i - \tilde{\nabla}\pi_i \otimes \tilde{\nabla}\rho^i . \tag{7}$$

The question of how to choose those local coordinates, which are Darboux coordinates for the cotangent bundle, remains open. The answer is not to be found in mathematics but in physics. In classical mechanics, the criterion for choosing a canonical momentum is provided by a Lagrangian; however, here, we do not have a Lagrangian. An alternative criterion more closely tailored to the framework presented here is provided by entropic dynamics [11]. From now on, we assume that the correct π_i coordinates have been identified.

The action of $\Omega[\cdot,\cdot]$ on two vectors $\bar{V} = d/d\lambda$ and $\bar{U} = d/d\mu$ is obtained using (6).

$$\tilde{\nabla}\rho^i(\bar{V}) = V^{1i} \quad \text{and} \quad \tilde{\nabla}\pi_i(\bar{V}) = V^{2i} . \tag{8}$$

The result is

$$\Omega[\bar{V}, \bar{U}] = V^{1i}U^{2i} - V^{2i}U^{1i} = \Omega_{\alpha i,\beta j} V^{\alpha i} U^{\beta j} \quad \text{where} \quad \Omega_{\alpha i,\beta j} = \begin{bmatrix} 0 & 1 \\ -1 & 0 \end{bmatrix} \delta_{ij} . \tag{9}$$

3.2. Hamilton's Equations and Poisson Brackets

Next, we derive the $2n$-dimensional $T^*\mathcal{S}^+$ analogues of the results that are standard in classical mechanics [14–16]. We seek those vector fields $\bar{V}(X)$ that generate flows (the congruence of integral curves) that preserve the symplectic structure in the sense that

$$\pounds_V \Omega = 0 , \tag{10}$$

where the Lie derivative [16] is

$$(\pounds_V \Omega)_{\alpha i,\beta j} = V^{\gamma k}\partial_{\gamma k}\Omega_{\alpha i,\beta j} + \Omega_{\gamma k,\beta j}\partial_{\alpha i} V^{\gamma k} + \Omega_{\alpha i,\gamma k}\partial_{\beta j} V^{\gamma k} . \tag{11}$$

Since, by Equation (9), the components $\Omega_{\alpha i,\beta j}$ are constant, $\partial_{\gamma k}\Omega_{\alpha i,\beta j} = 0$, we can rewrite $\pounds_V \Omega$ as

$$(\pounds_V \Omega)_{\alpha i,\beta j} = \partial_{\alpha i}(\Omega_{\gamma k,\beta j} V^{\gamma k}) - \partial_{\beta j}(\Omega_{\gamma k,\alpha i} V^{\gamma k}) , \tag{12}$$

which is the exterior derivative (roughly, the curl) of the covector $\Omega_{\gamma k,\alpha i}V^{\gamma k}$. By Poincare's lemma, requiring $\pounds_V\Omega = 0$ (a vanishing curl) implies that $\Omega_{\gamma k,\alpha i}V^{\gamma k}$ is the gradient of a scalar function, which we denote by $\tilde{V}(X)$,

$$\Omega_{\gamma k,\alpha i}V^{\gamma k} = \partial_{\alpha i}\tilde{V} \quad \text{or} \quad \Omega(\bar{V},\cdot) = \tilde{\nabla}\tilde{V}(\cdot)\,. \tag{13}$$

In the opposite direction, we can easily check that (13) implies $\pounds_V\Omega = 0$. Using (9), Equation (13) is more explicitly written as

$$\frac{d\rho^i}{d\lambda}\tilde{\nabla}\pi_i - \frac{d\pi_i}{d\lambda}\tilde{\nabla}\rho^i = \frac{\partial\tilde{V}}{\partial\rho^i}\tilde{\nabla}\rho^i + \frac{\partial\tilde{V}}{\partial\pi_i}\tilde{\nabla}\pi_i\,, \tag{14}$$

or

$$\frac{d\rho^i}{d\lambda} = \frac{\partial\tilde{V}}{\partial\pi_i} \quad \text{and} \quad \frac{d\pi_i}{d\lambda} = -\frac{\partial\tilde{V}}{\partial\rho^i}\,, \tag{15}$$

which we recognize as Hamilton's equations for a Hamiltonian function $\tilde{V}$. This justifies calling $\bar{V}$ the *Hamiltonian vector field* associated to the *Hamiltonian function* $\tilde{V}$. In other words, the flows that preserve the symplectic structure, $\pounds_V\Omega = 0$, are generated by Hamiltonian vector fields $\bar{V}$ associated to Hamiltonian functions $\tilde{V}$.

From (9) and (15) the action of the symplectic form Ω on two Hamiltonian vector fields $\bar{V} = d/d\lambda$ and $\bar{U} = d/d\mu$ generated, respectively, by $\tilde{V}$ and $\tilde{U}$, is

$$\Omega[\bar{V},\bar{U}] = \frac{d\rho^i}{d\lambda}\frac{d\pi_i}{d\mu} - \frac{d\pi_i}{d\lambda}\frac{d\rho^i}{d\mu} = \frac{\partial\tilde{V}}{\partial\rho^i}\frac{\delta\tilde{U}}{\delta\pi_i} - \frac{\partial\tilde{V}}{\partial\pi_i}\frac{\partial\tilde{U}}{\partial\rho^i} \overset{\text{def}}{=} \{\tilde{V},\tilde{U}\}\,, \tag{16}$$

where, on the right hand side, we have introduced the Poisson bracket notation. In other words, the action of Ω on two Hamiltonian vector fields is the Poisson bracket of the associated Hamiltonian functions. We can also check that the derivative of an arbitrary function $F(X)$ along the vector field $\bar{V} = d/d\lambda$ is

$$\frac{dF}{d\lambda} = \{F,\tilde{V}\}\,. \tag{17}$$

Thus, *the Hamiltonian formalism that is so familiar in physics emerges from purely geometrical considerations*. It might be desirable to adopt a more suggestive notation; instead of $(\tilde{V},\lambda)$ let us write $(\tilde{H},\tau)$. Then, the flow generated by a Hamiltonian function $\tilde{H}$ and parameterized by "time" τ is given by Hamilton's equations in the standard form,

$$\frac{d\rho^i}{d\tau} = \frac{\partial\tilde{H}}{\partial\pi_i} \quad \text{and} \quad \frac{d\pi_i}{d\tau} = -\frac{\partial\tilde{H}}{\partial\rho^i}\,, \tag{18}$$

and the τ evolution of any well-behaved function $f(X)$ is given by

$$\frac{df}{d\tau} = \bar{H}(f) = \{f,\tilde{H}\} \quad \text{with} \quad \bar{H} = \frac{\partial\tilde{H}}{\partial\pi_i}\frac{\partial}{\partial\rho^i} - \frac{\partial\tilde{H}}{\partial\rho^i}\frac{\partial}{\partial\pi_i}\,. \tag{19}$$

The difference with classical mechanics is that, here, the degrees of freedom are probabilities and not ontic variables such as, for example, the positions of particles.

3.3. The Normalization Constraint

Since our actual interest is not in flows on $T^*\mathcal{S}^+$ but on the bundle $T^*\mathcal{S}$ of normalized probabilities, we shall restrict ourselves to flows that preserve the normalization of probabilities. Let

$$|\rho| \overset{\text{def}}{=} \sum_{i=1}^{n}\rho^i \quad \text{and} \quad \tilde{N} \overset{\text{def}}{=} 1 - |\rho|\,. \tag{20}$$

We seek those special Hamiltonians $\tilde{H}$ such that the initial condition $\tilde{N} = 0$ is preserved by the flow, that is,

$$\partial_\tau \tilde{N} = \{\tilde{N}, \tilde{H}\} = 0 \quad \text{or} \quad \sum_i \frac{\partial \tilde{H}}{\partial \pi_i} = \sum_i \frac{d\rho^i}{d\tau} = 0 \,. \tag{21}$$

Indeed, the actual quantum Hamiltonians will preserve $\tilde{N} = const.$ even when the constant does not vanish [11]. Since the probabilities ρ^i must remain positive, we further require that $d\rho^i/d\tau \geq 0$ when $\rho^i = 0$.

We can also consider the Hamiltonian flow generated by $\tilde{N}$ and parameterized by v. From Equation (15) the corresponding Hamiltonian vector field $\bar{N}$ is given by

$$\bar{N} = N^{\alpha i} \frac{\partial}{\partial X^{\alpha i}} \quad \text{with} \quad N^{\alpha i} = \frac{dX^{\alpha i}}{dv} = \{X^{\alpha i}, \tilde{N}\} \,, \tag{22}$$

or, more explicitly,

$$N^{1i} = \frac{d\rho^i}{dv} = 0 \quad \text{and} \quad N^{2i} = \frac{d\pi_i}{dv} = 1 \,. \tag{23}$$

The integral curves generated by $\tilde{N}$ are found by integrating (23). The result is

$$\rho^i(v) = \rho^i(0) \quad \text{and} \quad \pi_i(v) = \pi_i(0) + v \,, \tag{24}$$

which amounts to shifting all momenta by the i-independent parameter v. We can also see that, if $\tilde{N}$ is conserved along $\tilde{H}$, then $\tilde{H}$ is conserved along $\bar{N}$.

$$\frac{d\tilde{H}}{dv} = \{\tilde{H}, \tilde{N}\} = 0 \,, \tag{25}$$

which implies that the conserved quantity $\tilde{N}$ is the generator of a symmetry transformation.

To summarize: the phase space of interest is $T^*\mathcal{S}$, but the description is simplified by using the unnormalized coordinates ρ of the larger embedding space $T^*\mathcal{S}^+$. The introduction of one superfluous ρ coordinate forces us to also introduce one superfluous π momentum. We eliminate the extra coordinate by imposing the constraint $\tilde{N} = 0$. We eliminate the extra momentum by declaring it unphysical; the shifted point $(\rho', \pi') = (\rho, \pi + v)$ is declared to be equivalent to (ρ, π). This equivalence is described as a global "gauge" symmetry which, as we shall see later in the paper, is the reason why quantum mechanical states are represented by rays rather than vectors in a Hilbert space.

4. The Information Geometry of E-Phase Space

Our next goal is to extend the metric of the simplex $\mathcal{S}$—given by information geometry—to the full e-phase space, $T^*\mathcal{S}$. The extension can be carried out in many ways [9–11,42]. The virtue of the derivation below is that the number of input assumptions is kept to a minimum.

4.1. The Metric on the Embedding E-Phase Space $T^*\mathcal{S}^+$

First, we assign a metric to the embedding bundle $T^*\mathcal{S}^+$; then, we consider the metric it induces on $T^*\mathcal{S}$. The metric of the space $\mathcal{S}^+$ of unnormalized probabilities [13,43] is

$$\delta\ell^2 = g_{ij}\delta\rho^i\delta\rho^j \quad \text{with} \quad g_{ij} = A(|\rho|)n_i n_j + \frac{B(|\rho|)}{2\rho^i}\delta_{ij} \,, \tag{26}$$

where n is a covector with components $n_i = 1$ for all $i = 1\ldots n$ and $A(|\rho|)$ and $B(|\rho|)$ are smooth scalar functions of $|\rho| = \sum \rho^i$. Since *the only tensor at our disposal is* g_{ij} the length element of $T^*\mathcal{S}^+$ must be of the form

$$\delta\tilde{\ell}^2 = \alpha g_{ij}\delta\rho^i\delta\rho^j + \beta g_i^j\delta\rho^i\delta\pi_j + \gamma g^{ij}\delta\pi_i\delta\pi_j \,, \tag{27}$$

where α, β and γ are constants. Since $\delta\rho^i$ and $\delta\pi_i$ are vectors and covectors, the requirement that $\delta\tilde{\ell}^2$ induce the same magnitudes $g_{ij}\delta\rho^i\delta\rho^j$ on $T\mathcal{S}_\rho^+$ and $g^{ij}\delta\pi_i\delta\pi_j$ on $T^*\mathcal{S}_\rho^+$, as given by information geometry, implies that $\alpha = \gamma = 1$. To fix β, let us consider a curve $\rho = \rho(\tau)$ and $\pi = \pi(\tau)$ on $T^*\mathcal{S}^+$ and its flow-reversed or τ-reversed curve given by $\rho'(\tau) = \rho(-\tau)$ and $\pi'(\tau) = -\pi(-\tau)$. We require that the speed $(d\tilde{\ell}/d\tau)^2$ remains invariant under flow-reversal. Since, under flow-reversal, the mixed $\rho\pi$ terms in (27) change sign, it follows that invariance implies that $\beta = 0$. We emphasize that imposing that the e-phase space be symmetric under flow-reversal does not amount to imposing *time*-reversal invariance; time-reversal violations might still be caused by interaction terms in the Hamiltonian. The resulting line element, *which has been designed to be fully determined by information geometry*, takes the form

$$\delta\tilde{\ell}^2 = G_{\alpha i,\beta j}\delta X^{\alpha i}\delta X^{\beta j} = g_{ij}\delta\rho^i\delta\rho^j + g^{ij}\delta\pi_i\delta\pi_j \,. \tag{28}$$

4.2. A Complex Structure for $T^*\mathcal{S}^+$

The metric tensor G and its inverse G^{-1} can be used to lower and raise indices. In particular, with G^{-1}, we can raise the first index of the symplectic form $\Omega_{\alpha i,\beta j}$ in Equation (9).

$$G^{\alpha i,\gamma k}\Omega_{\gamma k,\beta j} \stackrel{\text{def}}{=} -J^{\alpha i}{}_{\beta j} \,. \tag{29}$$

The tensor J has several important properties. These are most easily derived by writing G and Ω in block matrix form, i.e.,

$$G^{-1} = \begin{bmatrix} g^{-1} & 0 \\ 0 & g \end{bmatrix}, \quad \Omega = \begin{bmatrix} 0 & 1 \\ -1 & 0 \end{bmatrix}, \quad J = \begin{bmatrix} 0 & -g^{-1} \\ g & 0 \end{bmatrix}. \tag{30}$$

We can immediately check that $JJ = -1$, which shows that J is a square root of the negative identity matrix. Thus, J endows $T^*\mathcal{S}^+$ with a complex structure. *To summarize, in addition to the symplectic Ω and metric G structures, the cotangent bundle $T^*\mathcal{S}^+$ is also endowed with a complex structure J.* Such highly structured spaces are generically known as Kähler manifolds. Here, we deal with a special Kähler manifold where the space of ρs is a statistical manifold and the spaces of πs are flat cotangent spaces. However, ultimately, the geometry of $T^*\mathcal{S}^+$ is only of marginal interest; what matters is the geometry it induces on the e-phase space $T^*\mathcal{S}$ of normalized probabilities, to which we turn next.

4.3. The Metric Induced on the E-Phase Space $T^*\mathcal{S}$

As we saw above the e-phase space $T^*\mathcal{S}$ can be obtained from the space $T^*\mathcal{S}^+$ by the restriction $|\rho| = 1$ and by identifying the gauge equivalent points (ρ^i, π_i) and $(\rho^i, \pi_i + n_i\nu)$. Consider two neighboring points (ρ^i, π_i) and (ρ'^i, π_i') with $|\rho| = |\rho'| = 1$, the metric induced on $T^*\mathcal{S}$ is defined as the shortest $T^*\mathcal{S}^+$ distance between (ρ^i, π_i) and the points on the ray defined by (ρ'^i, π_i'). Since the $T^*\mathcal{S}^+$ distance between (ρ^i, π_i) and $(\rho^i + \delta\rho^i, \pi_i + \delta\pi_i + n_i\nu)$ is

$$\delta\tilde{\ell}^2(\nu) = g_{ij}\delta\rho^i\delta\rho^j + g^{ij}(\delta\pi_i + n_i\nu)(\delta\pi_j + n_j\nu) \,, \tag{31}$$

the metric on $T^*\mathcal{S}$ is defined by $\delta\tilde{s}^2 = \min_\nu \delta\tilde{\ell}^2$. Imposing $|\delta\rho| = 0$, the value of ν that minimizes (31) is $\nu = -\langle\delta\pi\rangle = -\sum_i \rho^i\delta\pi_i$. Therefore, the metric on $T^*\mathcal{S}$, which measures the distance between neighboring rays, is

$$\delta\tilde{s}^2 = \sum_{i=1}^n \left[\frac{B(1)}{2\rho^i}(\delta\rho^i)^2 + \frac{2\rho^i}{B(1)}(\delta\pi_i - \langle\delta\pi\rangle)^2 \right]. \tag{32}$$

From now on, we set $B(1) = 1$, which only amounts to a choice of units and has no effect on our results. (In [11], we chose $B(1) = \hbar$.)

Although the metric (32) is expressed in a notation that may be unfamiliar, it turns out to be equivalent to the well-known *Fubini–Study metric*. Thus, the recognition that the e-phase space is the cotangent bundle of a statistical manifold led us to a novel derivation based on information geometry.

An important feature of the $T^*\mathcal{S}$ metric (32) is that, except for the irrelevant constant $B(1)$, it has turned out to be independent of the particular choices of the *functions* $A(|\rho|)$ and $B(|\rho|)$ (see Equation (26)) that define the geometries of the embedding spaces $\mathcal{S}^+$ and $T^*\mathcal{S}^+$. Therefore, without any loss of generality, we can simplify the analysis considerably by choosing $A(|\rho|) = 0$ and $B(|\rho|) = 1$, which gives the embedding spaces the simplest possible geometries, namely, they are flat. With this choice the $T^*\mathcal{S}^+$ metric, Equation (28) becomes

$$\delta\tilde{\ell}^2 = \sum_{i=1}^{n}\left[\frac{1}{2\rho^i}\delta\rho_i^2 + 2\rho^i\delta\pi_i^2\right] = G_{\alpha i,\beta j}\delta X^{\alpha i}\delta X^{\beta j} \quad \text{with} \quad G_{\alpha i,\beta j} = \begin{bmatrix} \delta_{ij}/2\rho_i & 0 \\ 0 & 2\rho_i\delta_{ij} \end{bmatrix} \quad (33)$$

and the tensor J, Equation (30), which defines the complex structure, becomes

$$J^{\alpha i}{}_{\beta j} = -G^{\alpha i,\gamma k}\Omega_{\gamma k,\beta j} = \begin{bmatrix} 0 & -2\rho_i\delta_{ij} \\ \delta_{ij}/2\rho_i & 0 \end{bmatrix}. \quad (34)$$

4.4. Refining the Choice of Cotangent Space: Complex Coordinates

Having endowed the e-phase spaces $T^*\mathcal{S}^+$ and $T^*\mathcal{S}$ with both metric and complex structures, we can now revisit and refine our choice of cotangent spaces. So far, we assumed the cotangent space $T^*\mathcal{S}_\rho^+$ at ρ to be the flat Euclidean n-dimensional space R^n. It turns out that the cotangent space that is relevant to quantum mechanics requires a further restriction. To see what this is, we use the fact that $T^*\mathcal{S}^+$ is endowed with a complex structure, which suggests a coordinate transformation from (ρ, π) to complex coordinates $(\psi, i\psi^*)$,

$$\psi_i = \rho_i^{1/2}e^{i\pi_i} \quad \text{and} \quad i\psi_i^* = i\rho_i^{1/2}e^{-i\pi_i}, \quad (35)$$

Thus, a point $\psi \in T^*\mathcal{S}^+$ has coordinates

$$\psi^{\mu i} = \begin{pmatrix} \psi^{1i} \\ \psi^{2i} \end{pmatrix} = \begin{pmatrix} \psi_i \\ i\psi_i^* \end{pmatrix}, \quad (36)$$

where the index $\mu = 1,2$ takes two values (with $\mu, \nu, \ldots$ chosen from the middle of the Greek alphabet).

Since changing the phase $\pi_i \to \pi_i + 2\pi$ yields the same point ψ, we see that the new $T^*\mathcal{S}_\rho^+$ is a flat n-dimensional "hypercube" (its edges have a coordinate length of 2π) with the opposite faces identified (periodic boundary conditions). Thus, the new $T^*\mathcal{S}_\rho^+$ is locally isomorphic to the old R^n, which makes it a legitimate choice of cotangent space. (Strictly, $T^*\mathcal{S}_\rho^+$ is a parallelepiped; from (28), we see that the lengths of its edges are $\ell_i = 2\pi(2\rho_i)^{1/2}$ which vanish at the boundaries of the simplex.)

We can check that the transformation from real (ρ, π) to complex coordinates $(\psi, i\psi^*)$ is canonical, so that

$$\Omega_{\mu i,\nu j} = \begin{bmatrix} 0 & 1 \\ -1 & 0 \end{bmatrix}\delta_{ij}, \quad (37)$$

retains the same form as (9).

Expressed in ψ coordinates, the Hamiltonian flow generated by the normalization constraint (24) is the familiar phase shift $\psi_i(v) = \psi_i(0)e^{iv}$. Thus, the gauge symmetry induced by the constraint $\tilde{N} = 0$ is the familiar multiplication by a constant phase factor.

In ψ coordinates, the metric G on $T^*\mathcal{S}^+$ Equation (33) becomes

$$\delta\tilde{\ell}^2 = -2i\sum_{i=1}^{n}\delta\psi_i\delta i\psi_i^* = G_{\mu i,\nu j}\delta\psi^{\mu i}\delta\psi^{\nu j} \quad \text{where} \quad G_{\mu i,\nu j} = -i\delta_{ij}\begin{bmatrix} 0 & 1 \\ 1 & 0 \end{bmatrix}. \quad (38)$$

Finally, using the inverse $G^{\mu i,\lambda k}$ to raise the first index of $\Omega_{\lambda k,vj}$ gives the ψ components of the tensor J,

$$J^{\mu i}{}_{vj} = -G^{\mu i,\lambda k}\Omega_{\lambda k,vj} = \begin{bmatrix} i & 0 \\ 0 & -i \end{bmatrix}\delta_{ij} \,. \tag{39}$$

5. Hamilton–Killing Flows

In the previous sections we studied those Hamiltonian flows $\bar{K}$ that, in addition to preserving the symplectic form, are generated by a gauge invariant $\tilde{K}$ so they also preserve the normalization constraint $\tilde{N}$. Our next goal is to find those flows that also happen to preserve the metric G of $T^*\mathcal{S}^+$, that is, we want $\bar{K}$ to be a Killing vector. The vector field $\bar{K}$ is determined by the Killing equation [16], $\pounds_K G = 0$, or

$$(\pounds_K G)_{\mu i,vj} = K^{\lambda k}\partial_{\lambda k}G_{\mu i,vj} + G_{\lambda k,vj}\partial_{\mu i}K^{\lambda k} + G_{\mu i,\lambda k}\partial_{vj}K^{\lambda k} = 0 \,. \tag{40}$$

Since Equation (38) gives $\partial_{\lambda k}G_{\mu i,vj} = 0$, the Killing equation simplifies to

$$(\pounds_K G)_{\mu i,vj} = -i\begin{bmatrix} \frac{\partial K^{2j}}{\partial\psi_i} + \frac{\partial K^{2i}}{\partial\psi_j} & ; & \frac{\partial K^{1j}}{\partial\psi_i} + \frac{\partial K^{2i}}{\partial i\psi_j^*} \\ \frac{\partial K^{2j}}{\partial i\psi_i^*} + \frac{\partial K^{1i}}{\partial\psi_j} & ; & \frac{\partial K^{1j}}{\partial i\psi_i^*} + \frac{\partial K^{1i}}{\partial i\psi_j^*} \end{bmatrix} = 0 \,, \tag{41}$$

where $\partial/\partial i\psi_i^* \overset{\text{def}}{=} -i\partial/\partial\psi_i^*$. If we further require that $\bar{K}$ is a Hamiltonian flow, $\pounds_K\Omega = 0$, then $K^{\mu i}$ satisfies Hamilton's equations,

$$K^{1i} = \frac{\partial\tilde{K}}{\partial i\psi_i^*} \quad\text{and}\quad K^{2i} = -\frac{\partial\tilde{K}}{\partial\psi_i} \,. \tag{42}$$

Substituting into (41), we find

$$\frac{\partial^2\tilde{K}}{\partial\psi_i\partial\psi_j} = 0 \quad\text{and}\quad \frac{\partial^2\tilde{K}}{\partial\psi_i^*\partial\psi_j^*} = 0 \,. \tag{43}$$

Therefore, in order to generate a flow that preserves both G and Ω, the function $\tilde{K}(\psi,\psi^*)$ must be *linear* in both ψ and ψ^*.

$$\tilde{K}(\psi,\psi^*) = \sum_{i,j=1}^{n}\psi_i^*\hat{K}_{ij}\psi_j + \sum_{i=1}^{n}(\psi_i^*\hat{L}_i + \hat{M}_i\psi_i) + const \,. \tag{44}$$

The kernels $\hat{K}_{ij}$, $\hat{L}_i$ and $\hat{M}_i$ are independent of ψ and ψ^*. Imposing that the flow preserves the normalization constraint $\tilde{N} = const$, Equation (21), implies that $\tilde{K}$ must be invariant under the phase shift $\psi \to \psi e^{iv}$. Therefore, $\hat{L}_i = \hat{M}_i = 0$ and we conclude that

$$\tilde{K}(\psi,\psi^*) = \sum_{i,j=1}^{n}\psi_i^*\hat{K}_{ij}\psi_j + const \,. \tag{45}$$

The corresponding HK flow is given by Hamilton's equations

$$\frac{d\psi_i}{d\lambda} = K^{1i} = \frac{\partial\tilde{K}}{\partial i\psi_i^*} = \frac{1}{i}\sum_{j=1}^{n}\hat{K}_{ij}\psi_j \,, \tag{46}$$

$$\frac{di\psi_i^*}{d\lambda} = K^{2i} = -\frac{\partial\tilde{K}}{\partial\psi_i} = -\sum_{j=1}^{n}\psi_j^*\hat{K}_{ji} \,. \tag{47}$$

The constant in (45) can be dropped, because it has no effect on the flow. Taking the complex conjugate of (46) and comparing with (47) show that the kernel $\hat{K}_{ij}$ is Hermitian and that the corresponding Hamiltonian functionals $\tilde{K}$ are real.

$$\hat{K}_{ij}^* = \hat{K}_{ji} \quad \text{and} \quad \tilde{K}(\psi, \psi^*)^* = \tilde{K}(\psi, \psi^*) \,. \tag{48}$$

To summarize, *the preservation of the symplectic structure, the metric structure and the normalization constraint leads to Hamiltonian functions $\tilde{K}$ that are bilinear in ψ and ψ^**, Equation (45). This is the main result of this paper. To appreciate its significance, once again, we adopt a more suggestive notation, i.e., the flow generated by the Hamiltonian function

$$\tilde{H}(\psi, \psi^*) = \sum_{i,j=1}^n \psi_i^* \hat{H}_{ij} \psi_j \quad \text{is} \quad \frac{d\psi_i}{d\tau} = \{\psi_i, \tilde{H}\} \quad \text{or} \quad i\frac{d\psi_i}{d\tau} = \sum_{j=1}^n \hat{H}_{ij}\psi_j \,, \tag{49}$$

which is recognized as the Schrödinger equation. Beyond being Hermitian, the actual form of the kernel $\hat{H}_{ij}$ remains undetermined.

The central feature of Hamilton's Equations (46) or of the Schrödinger Equation (49) is that they are linear. Given two solutions $\psi^{(1)}$ and $\psi^{(2)}$ and arbitrary constants c_1 and c_2, the linear combination $\psi^{(3)} = c_1\psi^{(1)} + c_2\psi^{(2)}$ is also a solution and this is extremely useful in calculations. Unfortunately, this is an HK flow on the embedding space $T^*\mathcal{S}^+$ and, when the flow is projected onto the e-phase space $T^*\mathcal{S}$, the linearity is severely restricted by normalization. If $\psi^{(1)}$ and $\psi^{(2)}$ are normalized points on $T^*\mathcal{S}$, the superposition $\psi^{(3)}$ is not in general a normalized point on $T^*\mathcal{S}$, unless the constants c_1 and c_2 are chosen appropriately. Furthermore, the states $\psi'^{(1)} = \psi^{(1)}e^{iv_1}$ and $\psi'^{(2)} = \psi^{(2)}e^{iv_2}$ are supposed to be "physically" equivalent to the original $\psi^{(1)}$ and $\psi^{(2)}$, but, in general, the superposition $\psi'^{(3)} = c_1\psi'^{(1)} + c_2\psi'^{(2)}$ is not equivalent to $\psi^{(3)}$. In other words, *the mathematical linearity of (46) or (49) does not extend to a full-blown superposition principle for physically equivalent states*. On the other hand, any point ψ deserves to be called a "state" in the limited sense that it may serve as the initial condition for a curve in $T^*\mathcal{S}^+$. Since, given two states $\psi^{(1)}$ and $\psi^{(2)}$, their superposition $\psi^{(3)}$ is also a state, we see that the set of states $\{\psi\}$ forms a linear vector space. This is a structure that turns out to be very useful.

6. Hilbert Space

Above we saw that the possible initial conditions for an HK flow, the *points* ψ of $T^*\mathcal{S}^+$, form a linear vector space. To take full advantage of linearity we would like to endow this vector space with the additional structure of an inner product and turn it into a Hilbert space—a term which we loosely use to describe any complex vector space with a Hermitian inner product. The metric tensor G (Equation (38)) and the symplectic form Ω (Equation (37)) are supposed to act on *vectors* $d/d\lambda$; their action on the *points* ψ or (ρ, π) is not defined. However, the choice of inner product for the points ψ is natural, in the sense that the necessary ingredients, G and Ω, are already available.

We adopt the familiar Dirac notation to represent the states ψ as vectors $|\psi\rangle$. In order that the inner product $\langle\psi|\phi\rangle$ be preserved it is defined in terms of the preserved tensors G and Ω,

$$\langle\psi|\phi\rangle \stackrel{\text{def}}{=} \frac{1}{2}\left(G_{\mu i, \nu j} + \alpha\Omega_{\mu i, \nu j}\right)\psi^{\mu i}\phi^{\nu j} \,, \tag{50}$$

where α is a constant and, to follow convention, the overall constant is set to $1/2$. Using Equation (37) and (38), we obtain

$$\langle\psi|\phi\rangle = \frac{1}{2}(\psi_i, i\psi_i^*)(G + \alpha\Omega)\begin{pmatrix}\phi_j \\ i\phi_j^*\end{pmatrix} = \frac{1}{2}\sum_{i=1}^n\left((1 - i\alpha)\psi_i^*\phi_i + (1 + i\alpha)\phi_i^*\psi_i\right) \,. \tag{51}$$

To fix α, we impose that $\langle\psi|\phi\rangle^* = \langle\phi|\psi\rangle$, which implies that $\alpha = \pm i$. In order to comply with the standard convention that the inner product $\langle\psi|\phi\rangle$ is anti-linear in the first factor and linear in the second factor, we select $\alpha = +i$. The result is the familiar expression for the positive definite inner product,

$$\langle \psi | \phi \rangle \stackrel{\text{def}}{=} \frac{1}{2}\left(G_{\mu i, \nu j} + i\Omega_{\mu i, \nu j} \right) \psi^{\mu i}\phi^{\nu j} = \sum_{i=1}^{n} \psi_i^* \phi_i \, . \tag{52}$$

Here we see that the choice of $1/2$ as the overall constant leads to the standard relation $\langle \psi | \psi \rangle = |\rho|$. The map between points and vectors, $\psi \leftrightarrow |\psi\rangle$, is defined by $|\psi\rangle = \sum_i |i\rangle \psi_i$, where $\psi_i = \langle i | \psi \rangle$ and the vectors $\{|i\rangle\}$ form a basis that is orthogonal and complete.

The bilinear Hamilton function $\tilde{K}(\psi, \psi^*)$ with kernel $\hat{K}_{ij}$ can now be written as the expected value, $\tilde{K}(\psi, \psi^*) = \langle \psi | \hat{K} | \psi \rangle$, of the Hamiltonian operator $\hat{K}$ with matrix elements $\hat{K}_{ij} = \langle i | \hat{K} | j \rangle$. The corresponding HK flows are given by

$$i\frac{d}{d\lambda}\langle i | \psi \rangle = \langle i | \hat{K} | \psi \rangle \quad \text{or} \quad i\frac{d}{d\lambda}|\psi\rangle = \hat{K}|\psi\rangle \, , \tag{53}$$

which are described by unitary transformations $|\psi(\lambda)\rangle = \hat{U}_K(\lambda)|\psi(0)\rangle$ where $\hat{U}_K(\lambda) = \exp(-i\hat{K}\lambda)$. Finally, the Poisson bracket of two Hamiltonian functions $\tilde{U}[\psi, \psi^*]$ and $\tilde{V}[\psi, \psi^*]$ can be written in terms of the commutator of the associated operators, $\{\tilde{U}, \tilde{V}\} = -i\langle \psi | [\hat{U}, \hat{V}] | \psi \rangle$. Thus, the Poisson bracket is the expectation of the commutator. This *identity* is much sharper than Dirac's pioneering discovery that the quantum commutator of two quantum variables is *analogous* to the Poisson bracket of the corresponding classical variables.

7. Conclusions

There have been numerous attempts to derive or construct the mathematical formalism of quantum mechanics by adapting the symplectic geometry of classical mechanics. Such phase space methods invariably start from a classical phase space of positions and momenta (q^i, p_i) and, through some series of "quantization rules," posit a correspondence to self-adjoint operators $(\hat{Q}^i, \hat{P}_i)$ which no longer constitute a phase space. The connection to classical mechanics is lost. The interpretation of $\hat{Q}^i$ and $\hat{P}_i$ and even the answer to the question of what is ontic and what is epistemic become highly controversial. Probabilities play a secondary role in such formulations.

In this paper, we take a different starting point that places probabilities at the forefront. We discuss special families of curves—the Hamilton–Killing flows—that promise to be useful for the study of quantum mechanics. We show that the HK flows that preserve the symplectic and the metric structures of the e-phase space reproduce much of the mathematical formalism of quantum theory. It clarifies how the linearity of the Schrödinger equation, complex numbers and the Born rule $\rho_i = |\psi_i|^2$ (the Born rule for generic observables is discussed in [31,32]) follow from the symplectic and metric structures, while the normalization constraint leads to the equivalence of states along rays in a Hilbert vector space.

Acknowledgments: I would like to thank M. Abedi, D. Bartolomeo, C. Cafaro, N. Carrara, N. Caticha, F. X. Costa, S. DiFranzo, S. Ipek, D.T. Johnson, S. Nawaz, P. Pessoa, M. Reginatto and K. Vanslette, for valuable discussions and for their many insights and contributions at various stages of this program.

Conflicts of Interest: The authors declare no conflict of interest.

References

1. Bell, J. Against 'Measurement'. *Phys. World* **1990**, 33. [CrossRef]
2. Stapp, H.P. The Copenhagen Interpretation. *Am. J. Phys.* **1972**, *40*, 1098. [CrossRef]
3. Schlösshauer, M. Decoherence, the measurement problem, and interpretations of quantum mechanics. *Rev. Mod. Phys.* **2004**, *76*, 1267. [CrossRef]
4. Jaeger, G. *Entanglement, Information, and the Interpretation of Quantum Mechanics*; Springer: Berlin/Heidelberg, Germany, 2009.
5. Leifer, M.S. Is the Quantum State Real? An Extended Review of Ψ-ontology Theorems. *Quanta* **2014**, *3*, 67. [CrossRef]

6. Nelson, E. *Quantum Fluctuations*; Princeton UP: Princeton, NJ, USA, 1985.
7. Caticha, A. Entropic Dynamics, Time, and Quantum Theory. *J. Phys. A Math. Theor.* **2011**, *44*, 225303. [CrossRef]
8. Goyal, P.; Knuth, K.; Skilling, J. Origin of complex quantum amplitudes and Feynman's rules. *Phys. Rev. A* **2010**, *81*, 022109. [CrossRef]
9. Reginatto, M.; Hall, M.J.W. Quantum theory from the geometry of evolving probabilities. *AIP Conf. Proc.* **2012**, *1443*, 96.
10. Reginatto, M.; Hall, M.J.W. Information geometry, dynamics and discrete quantum mechanics. *AIP Conf. Proc.* **2013**, *1553*, 246.
11. Caticha, A. The Entropic Dynamics approach to Quantum Mechanics. *Entropy* **2019**, *21*, 943. [CrossRef]
12. Amari, S.; Nagaoka, H. *Methods of Information Geometry*; American Mathematical Society: Providence, RI, USA, 2000.
13. Caticha, A. Entropic Physics: Probability, Entropy, and the Foundations of Physics. Available online: Https://www.albany.edu/physics/faculty/ariel-caticha (accessed on 20 June 2021).
14. Arnold, V.I. *Mathematical Methods of Classical Mechanics*; Springer: Berlin/Heidelberg, Germany, 1997; Volume 60.
15. Souriau, J.-M. *Structure of Dynamical Systems—A Symplectic View of Physics*; Translation by Cushman-deVries, C.H.; Birkhäuser: Boston, MA, USA, 1997.
16. Schutz, B. *Geometrical Methods of Mathematical Physics*; Cambridge U.P.: Cambridge, UK, 1980.
17. Hermann, R. Remarks on the Geometric Nature of Quantum Phase Space. *J. Math. Phys.* **1965**, *6*, 1768. [CrossRef]
18. Kibble, T.W.B. Geometrization of Quantum Mechanics. *Commun. Math. Phys.* **1979**, *65*, 189–201. [CrossRef]
19. Heslot, A. Quantum mechanics as a classical theory. *Phys. Rev. D* **1985**, *31*, 1341. [CrossRef]
20. Anandan, J.; Aharonov, Y. Geometry of Quantum Evolution. *Phys. Rev. Lett.* **1990**, *65*, 1697. [CrossRef] [PubMed]
21. Cirelli, R.; Manià, A.; Pizzochero, L. Quantum mechanics as an infinite-dimensional Hamiltonian system with uncertainty structure: Parts I and II. *J. Math. Phys.* **1990**, *31*, 2891, 2898. [CrossRef]
22. Abe, S. Quantum-state space metric and correlations. *Phys. Rev. A* **1992**, *46*, 1667. [CrossRef] [PubMed]
23. Hughston, L.P. Geometric aspects of quantum mechanics. In *Twistor Theory*; Huggett, S.A., Ed.; Marcel Dekker: New York, NY, USA, 1995.
24. Ashtekar, A.; Schilling, T.A. Geometrical Formulation of Quantum Mechanics. In *On Einstein's Path*; Harvey, A., Ed.; Springer: New York, NY, USA, 1998.
25. de Gosson, M.A.; Hiley, B.J. Imprints of the Quantum World in Classical Mechanics. *Found. Phys.* **2011**, *41*, 1415. [CrossRef]
26. Elze, H.-T. Linear dynamics of quantum-classical hybrids. *Phys. Rev. A* **2012**, *85*, 052109. [CrossRef]
27. Wootters, W.K. Statistical distance and Hilbert space. *Phys. Rev. D* **1981**, *23*, 357. [CrossRef]
28. Brodie, D.J.; Hughston, L.P. Statistical Geometry in Quantum Mechanics. *Phil. Trans. R. Soc. Lond. A* **1998**, *454*, 2445. [CrossRef]
29. Goyal, P. From Information Geometry to Quantum Theory. *New J. Phys.* **2010**, *12*, 023012. [CrossRef]
30. Molitor, M. On the relation between geometrical quantum mechanics and information geometry. *J. Geom. Mech.* **2015**, *7*, 169. [CrossRef]
31. Johnson, D.T.; Caticha, A. Entropic dynamics and the quantum measurement problem. *AIP Conf. Proc.* **2012**, *1443*, 104.
32. Vanslette, K.; Caticha, A. Quantum measurement and weak values in entropic quantum dynamics. *AIP Conf. Proc.* **2017**, *1853*, 090003.
33. Nawaz, S.; Caticha, A. Momentum and uncertainty relations in the entropic approach to quantum theory. *AIP Conf. Proc.* **2012**, *1443*, 112.
34. Bartolomeo, D.; Caticha, A. Trading drift and fluctuations in entropic dynamics: Quantum dynamics as an emergent universality class. *J. Phys. Conf. Ser.* **2016**, *701*, 012009. [CrossRef]
35. Bartolomeo, D.; Caticha, A. Entropic Dynamics: The Schrödinger equation and its Bohmian limit. *AIP Conf. Proc.* **2016**, *1757*, 030002.
36. Demme, A.; Caticha, A. The Classical Limit of Entropic Quantum Dynamics. *AIP Conf. Proc.* **2017**, *1853*, 090001.
37. Nawaz, S.; Abedi, M.; Caticha, A. Entropic Dynamics on Curved Spaces. *AIP Conf. Proc.* **2016**, *1757*, 030004.
38. Ipek, S.; Caticha, A. Entropic quantization of scalar fields. *AIP Conf. Proc.* **2015**, *1641*, 345.
39. Ipek, S.; Abedi, M.; Caticha, A. Entropic Dynamics: Reconstructing Quantum Field Theory in Curved Spacetime. *Class. Quantum Grav.* **2019**, *36*, 205013. [CrossRef]
40. Ipek, S.; Caticha, A. The Entropic Dynamics of Quantum Scalar fields coupled to Gravity. *Symmetry* **2020**, *12*, 1324. [CrossRef]
41. Caticha, A.; Carrara, N. The Entropic Dynamics of Spin. *arXiv* **2020**, arXiv:2007.15719.
42. Caticha, A. Entropic Dynamics: Quantum Mechanics from Entropy and Information Geometry. *Ann. Physik* **2019**, *531*, 1700408. [CrossRef]
43. Campbell, L.L. An extended Čencov characterization of the information metric. *Proc. Am. Math. Soc.* **1986**, *98*, 135.

Proceeding Paper

Statistical Mechanics of Unconfined Systems: Challenges and Lessons [†]

Bruno Arderucio Costa [1,*] **and Pedro Pessoa** [2]

1   Instituto de Física Teórica, Universidade Estadual Paulista, São Paulo, SP 01140-070, Brazil
2   Department of Physics, University at Albany (SUNY), Albany, NY 12222, USA; ppessoa@albany.edu
*   Correspondence: bruno.arderucio@unesp.br
†   Presented at the 40th International Workshop on Bayesian Inference and Maximum Entropy Methods in Science and Engineering, online, 4–9 July 2021.

Abstract: Motivated by applications of statistical mechanics in which the system of interest is spatially unconfined, we present an exact solution to the maximum entropy problem for assigning a stationary probability distribution on the phase space of an unconfined ideal gas in an anti-de Sitter background. Notwithstanding the gas' freedom to move in an infinite volume, we establish necessary conditions for the stationary probability distribution solving a general maximum entropy problem to be normalizable and obtain the resulting probability for a particular choice of constraints. As a part of our analysis, we develop a novel method for identifying dynamical constraints based on local measurements. With no appeal to a priori information about globally defined conserved quantities, it is therefore applicable to a much wider range of problems.

Keywords: maximum entropy; unconfined gases; general relativity; anti-de Sitter spacetime

Citation: Costa, B.A.; Pessoa, P. Statistical Mechanics of Unconfined Systems: Challenges and Lessons. *Phys. Sci. Forum* **2021**, *3*, 8. https://doi.org/10.3390/psf2021003008

Academic Editors: Wolfgang von der Linden and Sascha Ranftl

Published: 9 December 2021

Publisher's Note: MDPI stays neutral with regard to jurisdictional claims in published maps and institutional affiliations.

1. Introduction

Despite its overarching success, the statistical mechanics of Gibbs [1] falls short in its description of spatially unconfined systems. In the canonical ensemble, a classical gas of N particles is described by a Gibbs probability distribution $f(y)$, $y \in \mathcal{Y}$ defined over the phase space $\mathcal{Y} = \{(q_i, p_i)\}$, for $i \in \{1, 2, \ldots, N\}$ where q_i and p_i are the position and momentum for the i-th particle). The Gibbs distribution is of the form $f(y) \propto e^{-\beta H(y)}$, where β is the unit-corrected inverse temperature (throughout this paper, we choose units in which $k_B = c = 1$) and H is the system Hamiltonian. In cases where the gas is unconfined, meaning the set of allowed positions is unlimited, the canonical ensemble may not lead to a valid probability distribution. The normalization condition in this ensemble is

$$\int dy\, e^{-\beta H(y)} < \infty. \tag{1}$$

One usually can write the Hamiltonian in the block-diagonal form

$$H(y) = \sum_{ij} F(q_{ij}) + \sum_i \frac{p_i^2}{2m},$$

where $q_{ij} = q_i - q_j$. Condition (1) is manifestly not satisfied whenever F falls off to infinity slower or as slow as $1/|q_{ij}|$. In particular, the normalization condition is not met for Newton/Coulomb-type interactions or a constant in q_{ij} (i.e., the ideal gas).

In the kinetic theory, the situation is no better. As already pointed out in ref. [2], solutions to the equilibrium Boltzmann equation are spatially homogeneous in the absence of external forces.

There are, however, examples of unconfined systems of physical interest, for example, an unconfined gas of particles interacting through gravity, such as a star or a galaxy cluster.

The statistical mechanics of such a system has been studied before and, to the best of our knowledge, every study thus far on the self-gravitating gas (or a gas of Coulomb-interacting particles [3]) requires confinement demanding that the particles are only allowed to occupy a limited set of position coordinates, i.e., confined in a box [4–7]. This approach allows one to bypass the normalization problem by restraining the integral (1) over the coordinates q_i to a compact domain of volume V.

Despite its effectiveness for astrophysical applications, progress in this direction leaves fundamental questions about the awkwardness of dealing with unconfined systems unanswered. For example, whether or not it is possible to display a system that reaches thermodynamical equilibrium while kinematically allowed to occupy an infinite volume. This article answers this question affirmatively.

2. Background

The work of Jaynes [8] solidified the statistical mechanics of Gibbs through what is currently known as the method of maximum entropy (MaxEnt). Further developments—starting with Shore and Johnson [9]—present MaxEnt as a method to select and update probability distributions f when information about the system is revealed by maximizing the functional

$$S[f] = -\int dy\, f(y) \log\left(\frac{f(y)}{\varphi(y)}\right),\tag{2}$$

where φ is a prior distribution, with constraints meant to represent the information at hand. Usually these constraints are in the form of expected values for a series of functions $a_\mu(y)$, namely sufficient statistics. The f that maximizes (2) under normalization, $\int dy\, f(y) = 1$, and a set of n expected value constraints, $A_\alpha = \int dy\, f(y) a_\alpha(y)$ for $\alpha \in \{1, \dots, n\}$, is the Gibbs distribution

$$f(y|\beta) = \frac{\varphi(y)}{Z} \exp\left\{ -\sum_\alpha \beta_\alpha a_\alpha(y) \right\},\tag{3}$$

where $Z(\beta)$ is a normalization factor, $Z = \int dy\, q(y) \exp\{-\sum_\alpha \beta_\alpha a_\alpha(y)\}$, and β_α are the Lagrange multipliers associated with the expected value constraint A_α. Fundamental factors in statistics imply that only entropies of the form (2) are appropriate for updating information—see, e.g., [9–11]. It follows from (3) that the Lagrange multipliers and the expected values are related through $A_\alpha = -\frac{\partial}{\partial \beta_\alpha} \log Z$.

The canonical ensemble of statistical mechanics assumes a system of interest connected to a heat bath, namely a much larger system that exchange energy with the system of interest (see, e.g., [10]). In this model, the only sufficient statistic is given by the Hamiltonian of the system $a_1(y) = H(y)$.

In the description given by the canonical ensemble, one may say that when the variational problem of extremizing (2) with the expected value of energy constraint cannot be solved, it means the system does not reach thermal equilibrium. This is the case for the self-gravitating gas, as explained in the introduction. It then follows that, in order to properly describe a thermodynamical system held to itself by gravity, we should evolve MaxEnt into the dynamical description, allowing the system to evolve in time.

Describing time evolution, in our context, means studying how the function f changes when a particle moves along its worldline (i.e., its trajectory in spacetime). Since we are dealing with massive particles, we can always use the proper time τ to parametrize their worldlines. Please note that a change in the position x^a is accompanied by a change in the tangent vectors to the trajectory, which are proportional to the momentum p^a. To keep our notation as simple as possible, we implicitly interpret f below as a function of τ using the composition $f(x^a(\tau), p^a(\tau))$.

In statistical mechanics, a very common choice for the prior $\varphi(y)$ in Equation (2) is a uniform distribution over $\mathcal{Y}$. This choice implements a symmetrical ignorance about the system: in the lack of information favouring a point over any other, it is fair to demand the prior to reflect the homogeneity. Under the understanding that time evolution preserves information, we have, adopting a uniform prior,

$$\frac{d}{d\tau}S[f(x^a, p^a)] = \frac{\delta S}{\delta f} \times \frac{df(x^a, p^a)}{d\tau}. \tag{4}$$

The first factor is, by Equation (2), nonzero. Hence, $\frac{dS}{d\tau} = 0$, implying $\frac{df}{d\tau} = 0$ and consequently [12],

$$\left(\frac{1}{m}p^\mu \frac{\partial}{\partial x^\mu} + \frac{dp^\mu}{d\tau}\frac{\partial}{\partial p^\mu} \right)f = 0, \tag{5}$$

where τ is the proper time along the integral curves of p^a and m is the mass of a particle of the gas.

Equation (5) provides us with a necessary condition the function f has to fulfil in order to extremize the entropy.

Our case of study is the anti-de Sitter spacetime, a maximally symmetric space with negative curvature. As for any space of constant curvature, its curvature tensor is fully determined by its scalar curvature R. In our case, $R = -12/\Lambda^2$, where Λ is the *anti-de Sitter length*. Consequently, the limit $\Lambda \to \infty$ recovers the local geometry of Minkowski spacetime. The anti-de Sitter space can be thought of as a solution for the vacuum Einstein's equation with a cosmological constant $-3/\Lambda^2$. For a lengthy discussion about this spacetime, we suggest the refs. [13,14]. For the purposes of the discussion presented here, note that although it is not globally hyperbolic, the anti-de Sitter spacetime can be foliated by a family of spacelike surfaces of infinite volume.

3. Solutions of the Boltzmann Equation in Anti-De Sitter Spacetime

We work with the universal covering space of an anti-de Sitter spacetime described by the metric

$$ds^2 = -\left(1 + \frac{r^2}{\Lambda^2}\right)dt^2 + \left(1 + \frac{r^2}{\Lambda^2}\right)^{-1}dr^2 + r^2 d\theta^2 + r^2 \sin^2\theta d\phi^2 , \tag{6}$$

with $t \in \mathbb{R}$, $r \in [0, \infty)$, $\theta \in [0, \pi]$ and $\phi \in [0, 2\pi)$. In these coordinates, it is clear that $\zeta^a = \left(\frac{\partial}{\partial t}\right)^a$ is a timelike Killing vector field.

The collisionless Boltzmann equation for an ideal gas in a generic curved spacetime follows from applying the geodesic equation to Equation (5) and reads [15]

$$\left(p^\mu \frac{\partial}{\partial x^\mu} - \Gamma^\mu_{\nu\rho}p^\nu p^\rho \frac{\partial}{\partial p^\mu} \right)f(x^\lambda, p^\lambda) = 0 , \tag{7}$$

where $\Gamma^\mu_{\nu\rho}$ are the Christoffel symbols. Equation (7) assumes the motion to be geodesic. This means that the ideal gas is an inbuilt assumption for its validity.

We shall seek spherically symmetric solutions of Equation (7) on the metric (6). For these solutions $f(x^\lambda, p^\lambda)$ can be written as a four-variable function $\tilde{f}(t, r, p^0, p^r)$, which, when substituted in (7), yields

$$p^0\frac{\partial\tilde{f}}{\partial t} + p^r\frac{\partial\tilde{f}}{\partial r} - \frac{2r}{r^2 + \Lambda^2}p^0 p^r \frac{\partial\tilde{f}}{\partial p^0} - \left[\frac{r(r^2 + \Lambda^2)}{\Lambda^4}(p^0)^2 - \frac{r}{r^2 + \Lambda^2}(p^r)^2\right]\frac{\partial\tilde{f}}{\partial p^r} = 0 , \tag{8}$$

restricted to the submanifold with $p^\theta = p^\phi = 0$.

Equation (8) admits a separation of variables as

$$\tilde{f}(t, r, p^0, p^r) = T(t)\mathcal{S}(r, p^0, p^r), \tag{9}$$

where T and $\mathcal{S}$ are functions to be determined.

The equation for t has a simple exponential solution $T(t) \propto e^{-t/t_c}$, while the equation for S cannot be solved using elementary methods unless the separation constant $1/t_c = 0$, in which case the solutions are referred to as *stationary*. It is important to bear in mind that the condition $df/d\tau = 0$, which follows directly from the maximization of the entropy, does not necessarily entail $df/dt = 0$.

By direct substitution into Equation (9), we can see that

$$
S = h\left(p^0\left(1 + \frac{r^2}{\Lambda^2}\right), -\left(1 + \frac{r^2}{\Lambda^2}\right)(p^0)^2 + \left(1 + \frac{r^2}{\Lambda^2}\right)^{-1}(p^r)^2 \right),
\tag{10}
$$

for any function $h \in \mathscr{C}^2(\mathbb{R}^2)$ solves Equation (8).

The arguments of the function h have a simple physical interpretation. The second is $g_{\mu\nu}p^\mu p^\nu$ when $p^\theta = p^\phi = 0$. This quantity is merely a constant, namely $-m^2$, where $m > 0$ is the mass of the particles constituting the gas. The first argument is an integral of the motion,

$$
\varepsilon \equiv -p^a \zeta_a,
\tag{11}
$$

the conserved energy of a particle [16].

This result is not surprising. Indeed, progress was made in the 1960s for simultaneously solving Einstein's equations and the equations of motion for matter consisting of a collection of particles. When a Hamiltonian can be defined for such systems, Liouville's equation is a consequence of Einstein's field equations. Fackerell [17] and Ehlers et al. [18], based on an analysis of the characteristic equations of (7), identified that the general solutions of Equation (5) can be written as functions of the conserved quantities. As a matter of fact, it follows that any function of first integrals of motion of the individual particles automatically satisfies the stationary Liouville Equation (5). This shows that not only is Equation (10) the general solution for the stationary spherically symmetric case, but also a particular solution for the general stationary case, i.e., when f is allowed to depend on the angular variables as well as having non-zero arguments p^θ and p^ϕ.

Please note that unlike in refs. [17,18], we are not interested in finding solutions to the full gravitational problem. Rather, we wish to preserve the analogy with the flat spacetime version of the problem, thereby neglecting any gravitational effects caused by the gas or its parts.

The nature of the constant $-m^2$ is different to the nature of ε. The latter is a first integral of the motion, i.e., a function of phase space that is a constant along the geodesics. The former is more conveniently thought of as a kinematic invariant. For simplicity, we henceforth focus on solutions with vanishing angular part of the four-momentum but we shall return to this point in due time. Hereafter, we use the $-m^2$ invariant to restrict S to the co-dimension one subspace $\mathcal{P}$ of the section of $\mathcal{Y}$ and write $S = \tilde{h}(\varepsilon)$ for a single-variable function $\tilde{h}$ of ε, defined in (11).

Naturally, not all solutions of the form $S = \tilde{h}(\varepsilon)$ are normalizable. The ubiquitous normalization constraint reads [19]

$$
\int_m^\infty dp^0 \int_0^\infty dr \, \frac{r^2}{\sqrt{1 + r^2/\Lambda^2}} \, \tilde{h}(\varepsilon(p^0, r)) < \infty,
\tag{12}
$$

leads to an asymptotic (large r) behaviour of $\tilde{h}$ falling off at least as quickly as

$$
\tilde{h}(\varepsilon) \sim \varepsilon^\alpha \quad \text{with} \quad \alpha < -1.
\tag{13}
$$

Equation (13) shows that normalizable solutions to the Liouville equation are achievable. Had we allowed for non-zero angular components of the four-momentum, this conclusion would not be challenged. Owing to the parity-reversal symmetry of the metric, we study the geodesics on the equatorial plane $\theta = \pi/2$ without loss of generality. Instead of Equation (10), we now have $S = h(\varepsilon, r^2 p^\phi, m^2)$ since $\ell \equiv p^a \psi_a = r^2 p^\phi$ is another independent integral of the motion associated with the Killing field $\psi^a = (\partial/\partial\phi)^a$. The constant of the motion ℓ can be thought of as the angular momentum of a particle per unit mass. Using an entirely analogous reasoning, we construct the two-variable function $\tilde{h}(\varepsilon, \ell)$, which behave as $\tilde{h} \sim \varepsilon^\alpha \ell^\gamma$ for large values of the radial coordinate. The normalization conditions for p^0, p^ϕ, and r demand, respectively, that $\alpha < -1$, $\gamma < -1$, and $\alpha + \gamma < -2$, the last of which, of course, is redundant. This shows that there are normalizable solutions to the Liouville equation also when the angular part of the four-momentum is free to assume any kinematically allowed value. In other words, normalized distributions in the momenta are also automatically normalizable with respect to the configuration space variables.

4. Constraints and MaxEnt

Because the format of f as $\tilde{h}(\varepsilon, \ell)$ was derived from imposing $dS/d\tau = 0$, we have thus far deduced a necessary condition for f to extremize the entropy, but by no means sufficient. We now show that among solutions obeying Equation (13), there are extremes of the entropy subjected to physically relevant constraints.

We adapt the arguments from Caticha [10] to model a "measuring device" one can use to extract information about the gas and subsequently use the acquired information to set constraints upon the variational problem. The procedure we describe below uses local measurements to impose constraints on the system of interest. This idea can be used in other problems in general relativity where the use of global conditions are not as straightforward or uniquely defined as they are in non-relativistic mechanics. An idea similar to ours was discussed in flat spacetime by Kubli and Herrmann in ref. [20], where the authors elegantly described the interaction between their measuring device and the system of interest through an extra interaction term in the Hamiltonian. Our approach instead appeals to kinetic theory, is more general, and agrees with their results for relativistic particles in flat spacetime.

Our measuring device D consists of another ideal gas; this time confined into a box whose walls are perfectly permeable to the unconfined gas' particles, but impenetrable to the particles of D. As with any measuring device, D has to interact with the system of interest. We model this interaction by elastic collisions between particles of D and particles of the unconfined gas. D itself being an ideal gas, its constituents follow geodesics between successive collisions and hence cannot absorb energy from the gravitational field. In addition, we make the box large enough so that the thermodynamic limit can be applied to the gas D, but small in comparison to Λ^3 so that D possesses a well-defined total energy E. If this energy changes by an amount ΔE, a small subset I of the particles of the unconfined gas must have suffered a change of $\Delta \sum_{i \in I} \varepsilon_i = -\Delta E$. Therefore,

$$\mathcal{E} \equiv \sum_i \varepsilon_i + E \tag{14}$$

is an *unknown* constant, and the sum runs over all the particles of the unconfined gas. It is important to emphasize that $\mathcal{E}$ does not have the same status as the "total energy" in the non-relativistic theory. Its physical interpretation is neither as straightforward nor easily relatable to the Arnowitt-Deser-Misner conserved mass of the spacetime, which would include contributions of pure gravity, as well as the backreaction of the gas onto the metric.

With the knowledge of the value of E given by the measurement in D [21], we impose the normalization constraint and

$$\int_{\mathcal{P}} \{\mathcal{E} - \varepsilon(y)\} f(y) dy = E, \tag{15}$$

which leads to the Gibbs distribution

$$f(y) = \frac{1}{Z} e^{-\beta \varepsilon(y)} \propto e^{-\beta\, p^0\, (1 + r^2/\Lambda^2)}. \tag{16}$$

Here, for sake conformity with the notation from Section 2, we changed the notation from discrete sums $\sum_i$ to integrals $\int_{\mathcal{P}}$.

Clearly Equation (16) respects condition (13) and consequently the normalization constant Z is finite. A substitution of (16) into (15) reveals

$$E = \frac{\partial}{\partial \beta} \log Z + \mathcal{E}. \tag{17}$$

The partition function, as for any ideal gas, factors out [22] as $Z = \zeta^N$ and can be calculated explicitly:

$$\zeta = 4\pi \int_m^\infty dp^0 \int_0^\infty dr \; \frac{r^2 \exp\left[-\beta p^0 \left(1 + \frac{r^2}{\Lambda^2}\right)\right]}{\sqrt{1 + \frac{r^2}{\Lambda^2}}}$$
$$= \frac{2\pi\Lambda^3}{\beta} e^{-\beta m/2}\left[(1 + \beta m)K_0\left(\frac{\beta m}{2}\right) - \beta m K_1\left(\frac{\beta m}{2}\right)\right], \tag{18}$$

where $K_\nu(z)$ is the modified Bessel function of the second kind. The flat spacetime is recovered from Equation (18) in the limit $\Lambda \to \infty$, in which case the partition function diverges as it should.

Remarkably, substituting Equation (18) into (17), we see that $\mathcal{E}$ does not depend on the anti-de Sitter length Λ, suggesting that the equations of state so obtained could accurately describe the gas in flat space as well. A calculation of the entropy from Equations (16) and (18) yields a term independent of Λ and another depending on Λ logarithmically. The latter, in the case of N indistinguishable particles, can be made small in comparison with the former for sufficiently large N.

5. Discussion and Prospects

With the example of an ideal gas in anti-de Sitter spacetime, we hope to have elucidated that it is not the kinematical possibility to occupy an infinite volume that prevents a system from having a well-posed equilibrium probability distribution. We emphasize that the results of Section 3 are independent of the choice of constraints made in Section 4. Specifically, all normalizable stationary solutions of the MaxEnt problem supplied with whatever set of constraints must obey condition (13).

We interpret the divergence of the partition function of the ideal gas in flat spacetime as follows. When the phase space $\mathcal{Y}$ describes a collection of a fixed number of particles and the partition function factors out as $Z = \zeta^N$ as above (see also footnote [22]), if particles of arbitrarily low energies can visit any point of the q_i section of $\mathcal{Y}$, the spatial integral of the partition function is forced to sample the infinite volume in all its glory.

The convergence in anti-de Sitter space is explained after describing its radial timelike geodesics. They can be calculated directly from (6) requiring $g_{ab}p^a p^b = -m^2$ together with the condition that ε is a constant along them. The results are sinusoidal functions r of the proper time τ [23]. A more geometrical interpretation is that timelike geodesics are oscillatory: they continue indefinitely departing from a point and reconverging to another, never reaching infinity. Higher amplitudes are accompanied by higher values of ε so that the ζ-integral only samples the infinite spatial volume "weighed down" by an appropriate decreasing function. This view is endorsed by the realization that Equation (18) is divergent on the limit $m \to 0$: null geodesics do not share the oscillatory behaviour of their timelike counterparts. Rather, they reach the null infinity $\mathscr{I}$ regardless of their energy ε.

In this respect, one can think of the anti-de Sitter geometry acting on massive particles as an "external field" binding the system together. Let us digress and illustrate this analogy with an example: an unconfined, non-relativistic ideal gas in Minkowski spacetime interacting with a spherically symmetric harmonic well is described by the Hamiltonian $H = \sum_i p_i^2/2m + 1/2\, m\omega^2 r_i^2$, where r_i is the radial distance from the i-particle and the centre of the well. For this system, the sum of the energies of each individual particle *is* the total energy and the canonical ensemble exists. In this ensemble, the probability distribution has the same Gaussian dependence on the radial variable as in Equation (16). A direct integration gives

$$\zeta = \frac{8\pi^3}{\beta^3 \omega^3}.\tag{19}$$

In this simple system, the frequency ω acts like the reciprocal of the anti-de Sitter length Λ. The expected value of the total energy per particle, $-\frac{\partial}{\partial\beta}\log\zeta = 3/\beta$, does not depend on ω, in direct analogy with our earlier observation that for the gas in anti-de Sitter, $\mathcal{E}$ is independent of Λ.

The static patch of the de Sitter spacetime permits another straightforward test case. This patch is described by the metric (6) under the replacement $\Lambda^2 \to -\Lambda^2$. The separation of variables for the dynamical case discussed in the paragraph below Equation (8) applies equally well for the de Sitter case. However, the static patch, being part of the steady-state universe, drives a family of geodesic-following particles apart from one another. Therefore, based on the interpretation above, we do not anticipate the existence of stationary normalizable solutions whatsoever. This is indeed the case. The convergence of the spatial integral of $\tilde{h}(\varepsilon)$ as r approaches Λ requires $\tilde{h} \sim \varepsilon^\alpha$ with $\alpha \geq 1/2$, while the convergence condition on the integral over p^0 requires $\alpha < -1$. The impossibility of satisfying both these conditions at once proves that no function $\tilde{h}$ can be normalized. Again, this conclusion would not be altered if we allow the angular momentum ℓ to be free. This can be seen immediately from the realization that an extra factor of ℓ^γ is regular on $r = \Lambda$ for all γ, therefore playing no role when studying the possible divergences near the de Sitter horizon at $r = \Lambda$.

In summary, this paper undertook two missions. First, it established general necessary conditions for the existence of normalizable stationary solutions for the maximum entropy problem for a relativistic ideal gas in anti-de Sitter spacetime. Such conditions show that normalizable solutions may exist even when there is no "wall" or external forces confining a gas in place. Second, it proposed a type of constraints based on local measurements that can have applications transcending the study of unconfined systems. Making use of these constraints, we found an explicit solution to the MaxEnt problem that is normalizable, as shown for the finiteness of ζ in (18).

The success of our programme for ideal gases invites its application for long-range interacting gases. It would be instructive to compare the dependence of thermodynamic properties of, say the self-gravitating star, on the "regularization procedure". That is to say, if the equations of state of a self-gravitating star in anti-de Sitter spacetime with very large Λ agree with, e.g., the large-V results from refs. [6,7]. We anticipate, of course, that the results thereby obtained will not agree with the ones from refs. [4,5] because these references introduce an interaction cut-off, which ultimately leads to extensivity, a property not expected to hold when the interaction is Newtonian until infinity [24]. Even more importantly in future applications, special attention must be given to any thermodynamic potential that turns out to be independent of Λ.

Author Contributions: Conceptualization, B.A.C. and P.P.; formal analysis, B.A.C.; methodology, B.A.C. and P.P.; validation, P.P.; writing—original draft preparation, review & editing, B.A.C. and P.P. All authors have read and agreed to the published version of the manuscript.

Funding: BAC was funded by Conselho Nacional de Desenvolvimento Científico e Tecnológico (CNPq), grant number 162288/2020-4.

Institutional Review Board Statement: Not applicable.

Informed Consent Statement: Not applicable.

Data Availability Statement: Not applicable.

Acknowledgments: The authors wish to thank Ariel Caticha for insightful conversations.

Conflicts of Interest: The authors declare no conflict of interest.

Abbreviations

The following abbreviations are used in this manuscript:
MaxEnt   MAXimum ENTropy principle

References and Notes

1. Gibbs, J. *Elementary Principles in Statistical Mechanics*; Yale University Press: New Haven, CT, USA, 1902; reprinted by Ox Bow Press: Woodbridge, CT, USA, 1981.
2. Pitaevskii, L.; Lifshitz, E. *Physical Kinetics: Course of Theoretical Physics Volume 10*; Number v. 10; Elsevier Science: Oxford, UK, 1981.
3. Davis, S.; Jain, J.; Bora, B. Computational statistical mechanics of a confined, three-dimensional Coulomb gas. *Phys. Rev. E* **2020**, *102*, 042137. [CrossRef] [PubMed]
4. Ahmad, F.; Saslaw, W.C.; Bhat, N.I. Statistical Mechanics of the Cosmological Many-Body Problem. *Astrophys. J.* **2002**, *571*, 576–584. [CrossRef]
5. Ahmad, F.; Saslaw, W.C.; Malik, M.A. Statistical Mechanics of the Cosmological Many-Body Problem. II. Results of Higher Order Contributions. *Astrophys. J.* **2006**, *645*, 940–949. [CrossRef]
6. de Vega, H.; Sánchez, N. Statistical mechanics of the self-gravitating gas: I. Thermodynamic limit and phase diagrams. *Nucl. Phys. B* **2002**, *625*, 409–459. [CrossRef]
7. de Vega, H.; Sánchez, N. Statistical mechanics of the self-gravitating gas: II. Local physical magnitudes and fractal structures. *Nucl. Phys. B* **2002**, *625*, 460–494. [CrossRef]
8. Jaynes, E.T. Information theory and statistical mechanics: I. *Phys. Rev.* **1957**, *106*, 620. [CrossRef]
9. Shore, J.; Johnson, R. Axiomatic derivation of the principle of maximum entropy and the principle of minimum cross-entropy. *IEEE Trans. Inf. Theory* **1980**, *26*, 26–37. [CrossRef]
10. Caticha, A. Entropic Physics: Probability, Entropy, and the Foundations of Physics. 2012. Available online: https://www.albany.edu/physics/faculty/ariel-caticha (accessed on 10 February 2021).
11. Vanslette, K. Entropic Updating of Probabilities and Density Matrices. *Entropy* **2017**, *19*, 664. [CrossRef]
12. Arnold, V. *Mathematical Methods of Classical Mechanics*; Graduate Texts in Mathematics; Springer: New York, NY, USA, 1978.
13. Hawking, S.; Ellis, G. *The Large Scale Structure of Space-Time*; Cambridge Monographs on Mathematical Physics; Cambridge University Press: Cambridge, UK, 1973.
14. Misner, C.; Thorne, K.; Thorne, K.; Wheeler, J.; W. H. Freeman and Company. *Gravitation*; Number pt. 3 in Gravitation; W. H. Freeman: New York, NY, USA, 1973.
15. Ellis, G.; Matravers, D.; Treciokas, R. Anisotropic solutions of the Einstein-Boltzmann equations: I. General formalism. *Ann. Phys.* **1983**, *150*, 455–486. [CrossRef]
16. Wald, R. *General Relativity*; University of Chicago Press: Chicago, IL, USA, 1984.
17. Fackerell, E.D. Relativistic Stellar Dynamics. *Astrophys. J.* **1968**, *153*, 643. [CrossRef]
18. Ehlers, J.; Geren, P.; Sachs, R.K. Isotropic solutions of the Einstein-Liouville equations. *J. Math. Phys.* **1968**, *9*, 1344–1349. [CrossRef]
19. Naturally, additional constraints may be present in the MaxEnt problem. The integral only ranges for non-negative values of p^0 because the condition that p^a is future-directed entails $p^0 > 0$.
20. Kubli, N.; Herrmann, H.J. Thermostat for a relativistic gas. *Phys. A Stat. Mech. Appl.* **2021**, *561*, 125261. [CrossRef]
21. This measurement can follow any procedure one would use to measure the total energy of an ideal gas confined in a box sitting on a desk. For example, one can put it on a set of weighing scales.
22. If one wishes to impose that the particles are indistinguishable from one another, the extra factor $1/N!$ is needed to account for the actual phase space being the quotient of all possible permutations of the pairs (q_i, p_i) in the product space of the one-particle phase spaces. The introduction of this factor is immaterial for our present discussion.
23. For reference, eliminating $\mathrm{d}t/\mathrm{d}\tau$ in favour of ε and substituting on the dispersion relation, we obtain $(\mathrm{d}r/\mathrm{d}\tau)^2 = \varepsilon^2 - 1 - r^2/\Lambda^2$. Taking a derivative of this expression with respect to τ we obtain the equation of motion of a simple harmonic oscillator, $\mathrm{d}^2 r/\mathrm{d}\tau^2 + \Lambda^{-2} r = 0$. We note that the frequency of oscillation is independent of the particle's energy.
24. Pessoa, P.; Costa, B.A. Comment on Tsallis, C. Black Hole Entropy: A Closer Look. Entropy 2020, 22, 17. *Entropy* **2020**, *22*, 1110. [CrossRef] [PubMed]

Proceeding Paper

On Two Measure-Theoretic Aspects of the Full Bayesian Significance Test for Precise Bayesian Hypothesis Testing [†]

Riko Kelter [‡]

Department of Mathematics, University of Siegen, 57072 Siegen, Germany; riko.kelter@uni-siegen.de

† Presented at the 40th International Workshop on Bayesian Inference and Maximum Entropy Methods in Science and Engineering, online, 4–9 July 2021.

‡ Current address: University of Siegen, Department of Mathematics, Walter-Flex-Street 3, 57072 Siegen, Germany.

Abstract: The Full Bayesian Significance Test (FBST) has been proposed as a convenient method to replace frequentist p-values for testing a precise hypothesis. Although the FBST enjoys various appealing properties, the purpose of this paper is to investigate two aspects of the FBST which are sometimes observed as measure-theoretic inconsistencies of the procedure and have not been discussed rigorously in the literature. First, the FBST uses the posterior density as a reference for judging the Bayesian statistical evidence against a precise hypothesis. However, under absolutely continuous prior distributions, the posterior density is defined only up to Lebesgue null sets which renders the reference criterion arbitrary. Second, the FBST statistical evidence seems to have no valid prior probability. It is shown that the former aspect can be circumvented by fixing a version of the posterior density before using the FBST, and the latter aspect is based on its measure-theoretic premises. An illustrative example demonstrates the two aspects and their solution. Together, the results in this paper show that both of the two aspects which are sometimes observed as measure-theoretic inconsistencies of the FBST are not tenable. The FBST thus provides a measure-theoretically coherent Bayesian alternative for testing a precise hypothesis.

Keywords: Full Bayesian Significance Test (FBST); statistical hypothesis testing; e-value; p-value

Citation: Kelter, R. On Two Measure-Theoretic Aspects of the Full Bayesian Significance Test. *Phys. Sci. Forum* **2021**, *3*, 10. https://doi.org/10.3390/psf2021003010

Academic Editors: Wolfgang von der Linden and Sascha Ranftl

Published: 17 December 2021

Publisher's Note: MDPI stays neutral with regard to jurisdictional claims in published maps and institutional affiliations.

1. Introduction

Statistical hypothesis testing is an important method in a broad range of sciences [1]. However, the recent problems with the validity of research results have been termed a scientific replication crisis [2,3], at the core of which lie some fundamental flaws in the statistical analysis of data [4]. Various papers have discussed the reproducibility of research and often the inadequate use of null hypothesis significance tests (NHST) substantiates a major cause of the replication crisis [5]. This holds in particular in the biomedical and cognitive sciences [6,7], where the p-value is the gold standard for quantifying the evidence against a precise null hypothesis.

Bayesian hypothesis testing has become increasingly popular in the biomedical and cognitive sciences due to the above problems [8–10]. It is well known that Bayesian data analysis solves some of the problems of NHST by allowing researchers to make use of optional stopping [11,12] and by simplifying the interpretation of censored data [13]. Together, these aspects are consequence of Bayesian inference being consistent with the likelihood principle [13]. An appealing proposal for a Bayesian test of a precise hypothesis is the Full Bayesian Significance Test (FBST), which has been applied in a wide range of domains [8,14–18]. The FBST advocates the e-value as a Bayesian replacement of the frequentist p-value for quantifying the statistical evidence against a precise hypothesis [19]. The FBST is a fully Bayesian procedure [19], accords with the likelihood principle [15], and enjoys attractive asymptotic properties [20] next to transformation invariance [16]. However, the FBST seems to suffer from two aspects which are studied in detail in this paper. First, the reference criterion in the FBST is only defined up to Lebesgue null sets, which seems to be make the evidential threshold arbitrary. Thus, it seems that the FBST statistical evidence, the *e*-value, lacks a calibration. Second, the statistical evidence in the FBST seems to have no prior probability, which contradicts common Bayesian reasoning. For other criticisms on the FBST see Ly & Wagenmakers [21] and for a more optimistic perspective Kelter [22]. In this paper it is shown that both aspects can be solved by fixing a version of the posterior distribution for statistical inference, and assigning one of two possible interpretations to the prior probability of the statistical evidence in the FBST. These aspects have not yet been discussed extensively in the literature and present a further justification of the FBST as an attractive replacement of frequentist *p*-values to remedy the ongoing problems with the replication of scientific results. The plan of the paper is as follows: The next section outlines the theory behind the FBST. After that, the two problematic aspects mentioned above are detailed and illustrated by an example from medical research. The following section elaborates on the problems and provides solutions to them. After that, a conclusion is provided.

2. The Full Bayesian Significance Test

This section outlines the theory behind the FBST. First, the required notation is introduced.

2.1. Notation

In contrast to the frequentist approach, in the Bayesian approach the parameter $\theta \in \Theta$ is modelled as a random variable, and the data $y \in \mathcal{Y}$ are fixed. Denote by Θ the parameter space and $\mathcal{G}$ as the σ-algebra on Θ, and let P_ϑ be the prior probability measure on $\mathcal{G}$, leading to the triple $(\Theta, \mathcal{G}, P_\vartheta)$. The observed sample is modelled by the random variable $Y : \Omega \to \mathcal{Y}$ which takes values in the measurable space $\mathcal{Y}$, where $\mathcal{Y}$ is endowed with a σ-algebra $\mathcal{B}$. The uncertainty in the data generating mechanism producing a sample $Y(\omega) = y$ for $\omega \in \Omega$ is modelled via the assumption of a statistical model $\mathcal{P} := \{P_\theta : \theta \in \Theta\}$ which is dominated by a σ-finite measure ν. In practice, ν often is the Lebesgue measure λ. The latter requirement guarantees the existence of Radon-Nikodým derivatives $dP_\theta/d\lambda = f(y|\theta)$. Let $(\Omega, \mathcal{A}, P^*)$ be the product space defined as $\Omega := \Theta \times \mathcal{Y}$, $\mathcal{A} := \mathcal{G} \times \mathcal{B}$ and P^* the product measure induced by the selection of P_ϑ and $\mathcal{P}$, where P_θ must be a measurable function on $\mathcal{B}$ for every y on $\mathcal{Y}$. Thus, P_ϑ is the marginal distribution of P^* with respect to the parameter θ, and the marginal distribution with respect to Y is the prior predictive $P^\vartheta(B) := \int_\Theta P_\theta(B)dP_\vartheta$ for any $B \in \mathcal{B}$. The parameter, as noted above, is modelled mathematically as a random variable $\vartheta : \Omega \to \Theta$. The resulting operational models from a Bayesian point of view are thus given as

1. the prior model $(\Theta, \mathcal{G}, P_\vartheta)$
2. the statistical model $\mathcal{P}$ on $(\mathcal{Y}, \mathcal{B})$, leading to $(\mathcal{Y}, \mathcal{B}, \{P_\theta : \theta \in \Theta\})$, and
3. the posterior model $(\Theta, \mathcal{G}, \{P_{\vartheta|Y} : Y \in \mathcal{Y}\})$

The existence of the posterior distribution $P_{\vartheta|Y}$ is guaranteed on Polish spaces [23] and inference about θ is conducted with respect to the posterior distribution $P_{\vartheta|Y}$ with density $p(\theta|y) := dP_{\vartheta|Y}/d\lambda$, which exists under the assumption that $P_\vartheta << \lambda$ where $<<$ denotes absolute-continuity of P_ϑ with respect to the measure λ.

2.2. Theory behind the Full Bayesian Significance Test (FBST)

The Full Bayesian Significance Test (FBST) was originally developed by Pereira and Stern [14] as an alternative to frequentist null hypothesis significance tests based on the *p*-value. It was created under the assumption that a significance test of a sharp hypothesis had to be conducted, where a sharp hypothesis refers to any submanifold of the parameter space of interest [20]. This includes, in particular, precise hypotheses like $H_0 : \theta = \theta_0$ for $\theta_0 \in \Theta$ [15]. The FBST assumes a standard parametric statistical model, where $\theta \in \Theta \subseteq \mathbb{R}^p$ is a (possibly vector-valued) parameter of interest, $f(y|\theta)$ is the density corresponding to the model distribution $P_{Y|\vartheta}$ and $p(\theta)$ is the prior density corresponding to the prior distribution P_ϑ, where we again assume a dominating measure ν to guarantee the existence of Radon-Nikodým densities. A hypothesis H makes the statement that the parameter θ lies in the corresponding null set Θ_H, where for simple (or precise) hypotheses $\Theta_H := \{\theta_0\}$, where θ_0 is the value specified in $H : \theta = \theta_0$. The *Full Bayesian Significance Test (FBST)* then defines two quantities: $ev(H)$, which is the *e*-value supporting (or in favour of) the hypothesis H, and $\overline{ev}(H)$, the *e*-value against H, also called the *Bayesian evidence value against H* [14]. First, the posterior *surprise function* $s(\theta)$ and its maximum s^* restricted to the null set Θ_H are introduced:

Definition 1 (Posterior surprise function). *The posterior surprise function $s(\theta)$ for a reference function $r : \Theta \to (\mathcal{T}, \mathcal{C})$ from Θ to a measurable space $(\mathcal{T}, \mathcal{C})$ is defined as*

$$s(\theta) := \frac{p(\theta|y)}{r(\theta)} \tag{1}$$

In the definition of the posterior surprise function $s(\theta)$, the denominator $r(\theta)$ serves as a reference density, and often the measurable space $(\mathcal{T}, \mathcal{C})$ is equal to $(\mathbb{R}^d, \mathcal{B}(\mathbb{R}^d))$. When the improper flat reference function $r(\theta) = 1$ is used, the surprise function becomes the posterior density $p(\theta|y)$. Otherwise, a weakly informative prior density can be used as a reference function, see Pereira and Stern [16]. Then,

$$s^* := s(\theta^*) = \sup_{\theta \in \Theta_H} s(\theta) \tag{2}$$

is defined as the supremum of the surprise function $s(\theta)$ over the null hypothesis support. For a precise null hypothesis, s^* is simply $s(\theta_0)$. Next, the tangential set is introduced:

Definition 2 (Tangential set). *The tangential set $\overline{T}(v)$ is defined as*

$$\overline{T}(v) := \Theta \setminus T(v) \tag{3}$$

where

$$T(v) := \{\theta \in \Theta | s(\theta) \leq v\} \tag{4}$$

Thus, $T(v)$ includes all parameter values $\theta \in \Theta$ which attain a surprise function value $s(\theta)$ smaller or equal to the threshold v. The tangential set $\overline{T}(v)$ is then the set complement and includes all parameter values $\theta \in \Theta$ which yield a surprise function value $s(\theta)$ larger than v. Fixing $v = s^*$ yields $\overline{T}(s^*)$, which is called the *tangential set to the hypothesis H*. This set $\overline{T}(s^*)$ contains the points θ of the parameter space Θ with higher surprise (or corroboration relative to the reference function $r(\theta)$) than the point θ_0 in the null set Θ_H. Then, the cumulative surprise function is introduced which is required to compute the *e*-value in the final step:

Definition 3 (Cumulative surprise function). *The map $W : \Theta \to [0, 1]$ given by*

$$W(v) := \int_{T(v)} p(\theta|y)d\theta \tag{5}$$

is called the complementary cumulative surprise function, and

$$\overline{W}(v) := 1 - W(v) \tag{6}$$

is called the cumulative surprise function.

Thus, the complementary cumulative surprise function $W(v)$ is the integral of the posterior density $p(\theta|y)$ over the set $T(v)$, and the cumulative surprise function $\overline{W}(v)$ is simply the integral of the posterior density over the tangential set $\overline{T}(\nu)$. The final step towards the e-value is to integrate the posterior density $p(\theta|y)$ over this set:

Definition 4 (e-value). *The e-value against a sharp null hypothesis $H_0 : \theta = \theta_0$ is defined as*

$$\overline{ev}(H_0) := \overline{W}(s^*) \tag{7}$$

and can be interpreted as the Bayesian evidence against H_0.

Clearly, $\overline{ev}(H_0) := \overline{W}(s^*)$ is the integral of the density $p(\theta|y)$ over the tangential set $T(s^*)$, which can be interpreted as the integral of the posterior density $p(\theta|y)$ over all parameter values θ which fulfill the condition $s(\theta) \geq s^*$. The e-value $ev(H_0)$ supporting H is obtained as $ev(H) := 1 - \overline{ev}(H_0)$ under $r(\theta) := 1$. Large values of $\overline{ev}(H_0)$ thus indicate that the hypothesis H traverses low-density regions (or equivalently, that the alternative hypothesis traverses high-density regions) so that the *evidence against H_0 is large*. For $r(\theta) \neq 1$ the argument is identical as H_0 traverses low posterior-surprise regions then.

For theoretical properties of the FBST and the e-value see Pereira and Stern [16] and Kelter [18]. The FBST then uses $ev(H)$ to reject H if $ev(H)$ is sufficiently small (or when $\overline{ev}(H)$ is large) [14,15].

3. On Two Aspects of the FBST

Now, this section demonstrates the two aspects briefly mentioned in the introduction based on an illustrative example.

3.1. The Reference Criterion

To illustrate the first problem, data of Rosenman et al. [24] of the Western Collaborative Group Study about coronary heart disease is used.

Example 1 (Coronary heart disease data). *The Western Collaborative Group Study began in 1960 with 3524 male volunteers who were 39 to 59 years old and free of heart disease as determined by electrocardiogram. After the initial screening, the study population dropped to 3154 because of various exclusions. Multiple endpoints were studied and average follow-up continued for 8.5 years with repeat examinations. As an illustrative example, suppose interest lies in testing for differences in systolic blood pressure between light smokers and heavy smokers. Thus, we test the hypothesis $H_0 : \delta = 0$ against the alternative $H_1 : \delta \neq 0$ where we classify participants with more than 5 cigarettes per day as heavy smokers. A Bayesian two-sample t-test using the model of Rouder et al. [25] is conducted, and the left plot in Figure 1 shows the results of the FBST using a flat reference function $r(\delta) := 1$. The model is parameterized in the effect size δ of Cohen [26], and the e-value $\overline{ev}(H_0)$ is given as $\overline{ev}(H_0) = 0.4362$, which equals the posterior probability mass visualized as the blue area in the left plot of Figure 1. Thus, 43.62% of the posterior probability indicate evidence against the null hypothesis, and the situation is inconclusive. The right plot in Figure 1 shows the result of the FBST when replacing the flat reference function $r(\delta) := 1$ with a Cauchy $C(0, \sqrt{2})$ density (note the different scaling on the y-axis), which is also used as the prior on δ in the two-sample t-test. In this case, the e-value $\overline{ev}(H_0) = 0.4367$ indicates a similarly inconclusive situation and changes the result barely.*

Now, the above example shows that calculation of the e-value is straightforward and universally applicable. However, the parameter space Θ is continuous in the example (the effect size $\delta \in \mathbb{R}$ is a continuous quantity) and any usual prior distribution P_θ assigned to θ is absolutely continuous with respect to the Lebesgue measure λ. It is well-known that the posterior distribution $P_{\theta|Y}$ is absolutely continuous with respect to the prior distribution [27], and thus any P_θ-null-set $N \subset \Theta$ with $P_\theta(N) = 0$ is also a $P_{\theta|Y}$-null-set with $P_{\theta|Y}(N) = 0$. Problematically, the set $\Theta_0 := \{\delta_0\} = \{0\}$ which is used in the precise null hypothesis $H_0 : \delta = 0$ is a P_θ-null-set under both the improper flat and Cauchy prior, as both of these are absolutely continuous with respect to the Lebesgue measure

λ, and submanifolds are Lebesgue-null-sets [28]. Thus, $\lambda(\{\delta_0\} = \lambda(\{0\}) = 0$ implies $P_\vartheta(\{0\}) = 0$ due to $P_\vartheta << \lambda$, which implies in turn that the posterior probability $P_{\vartheta|Y}(\{0\})$ of the value $\delta_0 = 0$ is a $P_{\vartheta|Y}$-null-set due to $P_{\vartheta|Y} << P_\vartheta$. As a consequence, the value of the posterior density $p(0|y) = 9.4693$ which is shown as the blue point in the left plot of Figure 1 could be chosen arbitrarily. Problematically, this value is used as the reference criterion in the calculation of the e-value $\overline{\mathrm{ev}}(H_0)$ in the computation of the tangential set $\overline{T}(\nu)$. Thus, one could assign $p(0|y)$ an entirely different value, say, $c \in \mathbb{R}$, and obtain a different e-value $\overline{\mathrm{ev}}(H_0)$ than the one calculated from the value $p(0|y) = 9.4693$. This seems to render the calculation of the statistical evidence $\overline{\mathrm{ev}}(H_0)$ in the FBST arbitrary, questioning the use of the procedure.

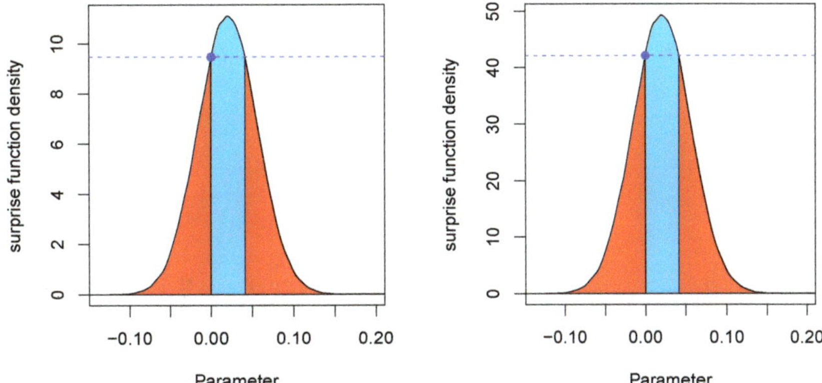

Figure 1. Results of the Full Bayesian Significance Test using a flat reference function (**left**) and a $C(0, \sqrt{2})$ Cauchy density as reference function (**right**) for testing the hypothesis of no difference $H_0 : \delta = 0$ in terms of systolic blood pressure between smokers and non-smokers.

3.2. Prior Probability of the e-Value

The second issue with the FBST may be phrased as the e-value having no valid prior probability. In fact, the e-value in Equation (7) is based on the cumulative surprise function $W(s^*)$, which itself depends on the tangential set $T(s^*)$ and the posterior density $p(\theta|y)$. Before data $y \in \mathcal{Y}$ are observed, the posterior $P_{\vartheta|Y}$ has not been realized as $P_{\vartheta|Y=y}$ and thus there exists no prior probability P_ϑ which is associated with the e-value. Even the tangential set $\overline{T}(s^*) := \{\theta \in \Theta | s(\theta) > s^*\}$ which is a subset of Θ seems to have no prior probability, because it depends on the surprise function $s(\theta)$ which itself depends on the posterior density $p(\theta|y)$, compare Equation (1). Thus, the statistical evidence in the FBST seems to escape the natural Bayesian transition from prior to posterior probability.

4. Solutions to the Two Aspects

4.1. The Reference Criterion

If the above criticism that the reference criterion in the FBST is arbitrary would hold, the procedure would be of little use in practice. However, the solution to the problem is given by fixing a specific version of the posterior distribution and performing all calculations conditional on fixing such a version. It is well known that probability distributions (which are probability measures corresponding to a random variable) are defined up to Lebesgue-null-sets (when they are dominated by the Lebesgue measure). The values on null-sets do not influence these probability measures and therefore they are identified with each other whenever they only differ on Lebesgue-null-sets [28]. Technically, this corresponds to the shift from the vector space $\mathcal{L}^p$

$$\mathcal{L}^p(\Omega, \mathcal{A}, \mu) := \left\{ f : \Omega \to \mathbb{K} \,\middle|\, f \text{ is measurable}, \int_\Omega |f(x)|^p d\mu(x) < \infty \right\} \tag{8}$$

on a probability space $(\Omega, \mathcal{A}, \mu)$, $\mathbb{K} \in \{\mathbb{R}, \mathbb{C}\}$ for $0 < p < \infty$ to the quotient space L^p, see Bauer [28]. The latter space is defined as $L^p := \mathcal{L}^p / \mathcal{N}$, where

$$\mathcal{N} := \left\{ f \in \mathcal{L}^p \middle| f = 0 \; \mu\text{-almost-everywhere} \right\} \tag{9}$$

and the elements in L^p are equivalence classes. Thus, two elements $[f], [g] \in L^p$ are equal if and only if they differ only on μ-null-sets, that is, $[f] - [g] \in \mathcal{N}$. Thus, the arbitrariness of the reference criterion in the FBST exists only unless a specific representant of the equivalence class, in which the posterior density $p(\theta|y)$ is located, is selected. In the context of Example 1, this implies that a specific version of the posterior density $p(\delta|y)$ needs to be chosen, which fixes the densities value on $\delta_0 = 0$ (and the other values $\delta \in \Theta$). Thus, setting $p(\delta_0|y) := p(0|y) := 9.4693$ explicitly by definition fixes one representant of the equivalence class of $P_{\theta|Y}$ and bypasses the problem that the reference threshold $p(\delta_0|y)$ in the FBST is arbitrary. Whenever the posterior is obtainable as a closed-form solution, that is, follows a well-known probability density $\tilde{P}_{\theta|Y}$ with Lebesgue-density $\tilde{p}(\theta|y)$, setting $p(\theta|y) := \tilde{p}(\theta|y)$ as the value of this known probability density $\tilde{p}$ for the posterior density p in the FBST *by definition* solves the first problem. Whenever numerical techniques like Markov-Chain-Monte-Carlo (MCMC) are used to produce the posterior, the resulting posterior distribution $P_{\theta|Y}^{MCMC}$ and the posterior density $p^{MCMC}(\theta|y)$ approximate the true posterior distribution $P_{\theta|Y}$ and the posterior Lebesgue-density $p(\theta|y)$. Thus, setting $p(\theta|y) := p^{MCMC}(\theta|y)$ by definition for a fixed numerical technique like MCMC with given random number generator seed fixes a version of the posterior density and renders the reference threshold in the FBST unique. In Example 1 this equals the choice of $p(\delta_0|y) := 9.4693$ by definition (as MCMC sampling was used), and $p(\delta|y) := p^{MCMC}(\delta|y)$ for all $\delta \in \mathbb{R}$. In summary, the above considerations provide the following result:

Theorem 1. *Let $s^* := s(\theta^*) = \sup\limits_{\theta \in \Theta_H} s(\theta)$ be the supremum of the surprise function in the Full Bayesian Significance Test, and $\mathcal{L}^p$ and L^p the corresponding vector spaces on $(\Theta, \mathcal{G}, P_{\theta|Y})$ with quotient space $L^p / \mathcal{N}$ for $\mathcal{N} := \{ f \in \mathcal{L}^p | f = 0 \; \mu\text{-almost-everywhere} \}$. Whenever $P_{\theta|Y}$ is a known probability distribution $\tilde{P}_{\theta|Y}$ with Lebesgue-density $\tilde{p}(\vartheta|Y)$, defining $p(\theta|y) := \tilde{p}(\theta|y)$ pointwise for all $\theta \in \Theta$ renders the e-value $\overline{ev}(H_0)$ against $H_0 : \theta = \theta_0$ for $\theta_0 \in \Theta$ well-defined and unique for the choice of $p(\theta|y)$.*

Proof. See Appendix A. $\square$

Note that when using numerical methods such as MCMC, ergodic theory ensures that $P_{\theta|Y}^{MCMC} \to P_{\theta|Y}$ in distribution and $p_{\theta|Y}^{MCMC} \to p_{\theta|Y}$, that is, the MCMC posterior density approximates the posterior Lebesgue-density pointwise with increasing precision for increasing number of MCMC samples [29]. Thus, fixing a version of the posterior, Theorem 1 extends also to situations where numerical techniques such as MCMC are required.

4.2. Prior Probability of the e-Value

The solution to the second problem is more involved and less technical. Conceptually, from the above line of thought it is immediate that under absolutely continuous priors P_θ with respect to the Lebesgue measure λ, the prior probability $P_\theta(\Theta_0)$ will be zero for any precise null hypothesis $H_0 := \Theta_0$ with $\Theta_0 := \{\theta_0\}$ for $\theta_0 \in \Theta$. The posterior $P_{\theta|Y}$ is absolutely continuous with respect to the prior P_θ, so $P_{\theta|Y}(\Theta_0) = 0$. Thus, it is simply not possible to use a natural Bayesian workflow which assigns positive probability mass to a Lebesgue-null-set Θ_0 whenever the statistician uses an absolutely continuous prior distribution P_θ with respect to λ. Traditional Bayesian hypothesis testing and model selection bypasses this inconvenience by introducing an arbitrary mixture prior structure $P_\theta := \varrho \mathbb{1}_{\Theta_0} + (1 - \varrho) \tilde{P}_\theta$ which assigns positive probability mass $\varrho > 0$ to the null set Θ_0, and distributes the rest of the probability mass $(1 - \varrho) \in [0, 1]$ by means of a probability distribution $\tilde{P}_\vartheta$ on the alternative hypothesis space $\Theta_1 = \Theta \setminus \Theta_0$. Early proposals of such a mixture prior structure include Jeffreys [30] and Haldane [31], see also Robert [29] and Kleijn [23]. Such a prior allows computation of a Bayes factor, and furthermore, the Bayes factor itself also has no prior probability which is naturally associated with it. Importantly, this mixture prior structure imposes a dichotomy between hypothesis testing and parameter estimation, because such a mixture prior structure is reasonable only from a hypothesis testing perspective. Whenever parameter estimation is the goal, the assignment of probability mass $\varrho > 0$ to a specific value is highly questionable and often contradicts reasonable a priori beliefs. In these cases, prior beliefs are expressed better through a prior which is absolutely continuous with respect to the Lebesgue measure λ.

The FBST avoids the introduction of such a mixture structure and thus allows for a unified prior elicitation which is coherent both from a Bayesian hypothesis testing and Bayesian parameter estimation stance. Importantly, the *e*-value is intended to be a Bayesian replacement of the frequentist *p*-value which measures the statistical discrepancy between the observed data to an assumed precise hypothesis. Thus, the *e*-value provides the Bayesian evidence against such a precise hypothesis. From a measure-theoretic point of view, every precise null hypothesis is assumed to be false and the FBST thus aligns with the empirical rationalism of Popper [32]. For the use of testing a precise hypothesis as an approximation of a small interval hypothesis see Berger [33], Rousseau [34], Rao & Lovric [35] as well as Kelter [36]: Often, the approximation of a small interval hypothesis via a precise point null hypothesis will be bad, and thus the *e*-value does not assign positive probability mass to such a precise null hypothesis. Instead, the FBST quantifies the discrepancy between the observed data and the hypothetical precise null value, while simultaneously implementing I.J. Good's principle of least surprise [37–39]. Note further that the mathematical introduction of positive prior probability $\varrho > 0$ to a precise value $\theta_0 \in \Theta$ when using a mixture prior does not render such a precise hypothesis $H_0 : \theta = \theta_0$ more realistic in practice.

Furthermore, next to its measure-theoretic premises, there exists another argument which weakens the criticism that there is no prior probability of the *e*-value: When a prior distribution P_ϑ is selected and no data $y \in \mathcal{Y}$ has been observed, the posterior distribution can be identified conceptually as the prior distribution. Thus, replacing the posterior density $p(\theta|y)$ with the λ-density $p(\theta)$ of the prior P_ϑ yields $s(\theta) := \frac{p(\theta)}{r(\theta)}$, which implies that the tangential set $\overline{T}(v) := \Theta \setminus T(v)$ for $T(v) := \{\theta \in \Theta | s(\theta) \leq v\}$ includes those parameter values $\theta \in \Theta$ for which $p(\theta)/r(\theta) > v$. Using the fact that $s^* = p(\theta_0)/r(\theta_0)$ for a precise hypothesis $H_0 : \theta = \theta_0$ then, yields $\overline{T}(v) = \{\theta \in \Theta | p(\theta)/r(\theta) > p(\theta_0)/r(\theta_0)\}$. Plugging this tangential set into Equation (6) yields the *e*-value

$$\overline{ev}(H_0) := \overline{W}(s^*) = \int_{\overline{T}(s^*)} p(\theta)d\theta$$

which is the integral of the prior density $p(\theta)$ over $\overline{T}(s^*)$. When the reference function $r(\theta)$ is chosen as a flat improper prior $r(\theta) := 1$, this becomes

$$\overline{ev}(H_0) = \int_{\{\theta \in \Theta | p(\theta) > p(\theta_0)\}} p(\theta)d\theta$$

which is the integral of the prior density $p(\theta)$ over all values which attain higher prior density values than the null value θ_0 in $H_0 : \theta = \theta_0$. Thus, the *e*-value in such a case quantifies the discrepancy of the precise hypothesis $H_0 : \theta = \theta_0$ with the prior beliefs P_ϑ. The above line of thought provide the following result:

Theorem 2. *Let $r(\theta) := 1$. In case no data $y \in \mathcal{Y}$ has been observed, the e-value quantifies the discrepancy between the precise hypothesis $H_0 := \Theta_0$ for $\Theta_0 := \{\theta_0\}$ and $\theta_0 \in \Theta$ and the prior distribution P_ϑ, that is,*

$$\overline{ev}(H_0) = P_\vartheta(\{\theta \in \Theta | p(\theta) > p(\theta_0)\}) \tag{10}$$

Proof. See Appendix A. $\square$

Whenever $r(\theta) \neq 1$, the interpretation is more complicated because such a reference function incorporates a surprise element into the tangential set, but the conclusions remain the same. The *e*-value then quantifies the discrepancy between the precise hypothesis and the prior surprise.

5. Discussion

The Full Bayesian Significance Test (FBST) has been proposed as a convenient method to replace frequentist *p*-values for testing a precise hypothesis [14–16]. Although the FBST enjoys various appealing properties [8,19,20,40], two aspects of the FBST are sometimes observed as measure-theoretic inconsistencies of the procedure and have not been discussed rigorously in the literature. First, the FBST uses the posterior density as a reference for judging the Bayesian statistical evidence against a precise hypothesis. However, under absolutely continuous prior distributions, the posterior density is defined only up to Lebesgue null sets which renders the reference criterion arbitrary. Second, the FBST statistical evidence seems to have no valid prior probability. In this paper, it was shown that the former problem can be circumvented by fixing a version of the posterior density before using the FBST. Theorem 1 demonstrated that then, the *e*-value is well-defined and unique after observing the data $y \in \mathcal{Y}$.

The latter aspect is based on the measure-theoretic premises of the FBST. As shown in this paper, the FBST avoids the use of a mixture prior structure which imposes a dichotomy between Bayesian hypothesis testing and parameter estimation. Thus, the FBST is compatible with absolutely continuous priors with respect to the Lebesgue measure λ (the Bayes factor, for example, is not). As a consequence, there exists no prior probability of the *e*-value and a precise hypothesis $H_0 : \theta = \theta_0$ under an absolutely continuous prior P_ϑ. Theorem 2 showed that even then, the *e*-value has a proper interpretation from a prior perspective: It quantifies the a priori discrepancy of the hypothesis H_0 with the prior beliefs which are expressed by P_ϑ whenever the reference function $r(\theta)$ is flat. When $r(\theta) \neq 1$, the interpretation is more difficult but the conclusion remains the same.

Together, the results in this paper show that both of the two aspects which are sometimes observed as measure-theoretic inconsistencies of the FBST are not tenable. The FBST thus provides a measure-theoretically coherent Bayesian alternative for testing a precise hypothesis.

Funding: This research received no external funding.

Institutional Review Board Statement: Not applicable.

Informed Consent Statement: Not applicable.

Data Availability Statement: The R code to recreate all analyses and plots can be found at the Open Science Foundation at https://osf.io/25vsw/?view_only=e1e243c1e2a44646969fb75cc4c34d57.

Conflicts of Interest: The authors declare no conflict of interest.

Abbreviations

The following abbreviations are used in this manuscript:

FBST   Full Bayesian Significance Test
NHST   Null Hypothesis Significance Testing

Appendix A

Proof of Theorem 1. From Definition 1 and Equation (2) it follows that the tangential set $\overline{T}(v) := \Theta \setminus T(v)$ becomes $\overline{T}(s^*) := \Theta \setminus T(s^*)$, which equals the set

$$
\{\theta \in \Theta : s(\theta) > s(\theta^*)\} = \left\{ \theta \in \Theta : \frac{p(\theta|y)}{r(\theta)} > \frac{p(\theta^*|y)}{r(\theta^*)} \right\}
$$
$$
= \left\{ \theta \in \Theta : \frac{p(\theta|y)}{r(\theta)} > \frac{p(\theta_0|y)}{r(\theta_0)} \right\} \tag{A1}
$$

where the first equality uses Definition 2 and the second equality uses $\theta^* = \theta_0$ for a precise hypothesis $H_0 : \theta = \theta_0$ for $\theta_0 \in \Theta$. By assumption, the posterior distribution $P_{\theta|Y}$ is known to take the form $\tilde{P}_{\theta|Y}$ with Lebesgue-density $\tilde{p}(\theta|y)$. Defining the posterior density $p : \Theta \to \mathbb{R}^d$ pointwise as $p(\theta|y) := \tilde{p}(\theta|y)$ implies that the value $p(\theta_0|y)$ is equal to $\tilde{p}(\theta_0|y)$. Thus, the tangential set $\overline{T}(s^*)$ in Equation (A1) is well-defined and unique for this fixed value $p(\theta_0|y) := \tilde{p}(\theta_0|y)$. From Definition 3 and Equation (7) it follows that the e-value $\overline{\mathrm{ev}}(H_0)$ is well-defined and unique for the choice of $p(\theta|y)$. $\square$

Proof of Theorem 2. Let P_θ be the prior distribution and $r(\theta) := 1$. Suppose no data $y \in \mathcal{Y}$ has been observed, then the posterior distribution $P_{\theta|Y}$ can be identified as the prior distribution P_θ. Thus, replacing the posterior density $p(\theta|y)$ with the λ-density $p(\theta)$ of the prior P_θ yields $s(\theta) := \frac{p(\theta)}{r(\theta)}$, which implies that the tangential set $\overline{T}(v) := \Theta \setminus T(v)$ for $T(v) := \{\theta \in \Theta | s(\theta) \le v\}$ includes the parameter values $\theta \in \Theta$ which fulfill the condition $p(\theta)/r(\theta) > v$. It follows that $s^* = p(\theta_0)/r(\theta_0)$ for a precise hypothesis $H_0 : \theta = \theta_0$, and this yields $\overline{T}(v) = \{\theta \in \Theta | p(\theta)/r(\theta) > p(\theta_0)/r(\theta_0)\}$ for the tangential set to H_0. Using the latter in Equation (6) yields the e-value

$$
\overline{\mathrm{ev}}(H_0) := \overline{W}(s^*) = \int_{\overline{T}(s^*)} p(\theta)d\theta
$$

which is the integral of the prior density $p(\theta)$ over $\overline{T}(s^*)$. By assumption, $r(\theta) := 1$, so this becomes

$$
\overline{\mathrm{ev}}(H_0) = \int_{\{\theta \in \Theta | p(\theta) > p(\theta_0)\}} p(\theta)d\theta = P_\theta(\{\theta \in \Theta | p(\theta) > p(\theta_0)\})
$$

which is the statement in Equation (10). $\square$

References

1. Gigerenzer, G. Mindless statistics. *J.-Socio-Econ.* **2004**, *33*, 587–606. [CrossRef]
2. Pashler, H.; Harris, C.R. Is the Replicability Crisis Overblown? Three Arguments Examined. *Perspect. Psychol. Sci.* **2012**, *7*, 531–536. [CrossRef]
3. Baker, M.; Penny, D. Is there a reproducibility crisis? *Nature* **2016**, *533*, 452–454. [CrossRef] [PubMed]
4. McElreath, R.; Smaldino, P.E. Replication, communication, and the population dynamics of scientific discovery. *PLoS ONE* **2015**, *10*, 1–16. [CrossRef] [PubMed]
5. Ioannidis, J.P.A. What Have We (Not) Learnt from Millions of Scientific Papers with p-Values? *Am. Stat.* **2019**, *73*, 20–25. [CrossRef]
6. Button, K.S.; Ioannidis, J.P.; Mokrysz, C.; Nosek, B.A.; Flint, J.; Robinson, E.S.; Munafò, M.R. Power failure: Why small sample size undermines the reliability of neuroscience. *Nat. Rev. Neurosci.* **2013**, *14*, 365–376. [CrossRef] [PubMed]
7. Kelter, R. Bayesian alternatives to null hypothesis significance testing in biomedical research: A non-technical introduction to Bayesian inference with JASP. *BMC Med. Res. Methodol.* **2020**, *20*, 1–12. [CrossRef]
8. Kelter, R. Analysis of Bayesian posterior significance and effect size indices for the two-sample t-test to support reproducible medical research. *BMC Med. Res. Methodol.* **2020**, *20*, 1–18. doi: 10.1186/s12874-020-00968-2. [CrossRef]
9. Kelter, R. Bayesian survival analysis in STAN for improved measuring of uncertainty in parameter estimates. *Meas. Interdiscip. Res. Perspect.* **2020**, *18*, 101–119. [CrossRef]
10. Wagenmakers, E.J.; Morey, R.D.; Lee, M.D. Bayesian Benefits for the Pragmatic Researcher. *Curr. Dir. Psychol. Sci.* **2016**, *25*, 169–176. [CrossRef]

11. Edwards, W.; Lindman, H.; Savage, L.J. Bayesian statistical inference for psychological research. *Psychol. Rev.* **1963**, *70*, 193–242. [CrossRef]
12. Hendriksen, A.; de Heide, R.; Grünwald, P. Optional Stopping with Bayes Factors: A Categorization and Extension of Folklore Results, with an Application to Invariant Situations. *Bayesian Anal.* **2020**, in press, [CrossRef]
13. Berger, J.; Wolpert, R.L. *The Likelihood Principle*; Institute of Mathematical Statistics: Hayward, CA, USA, 1988.
14. Pereira, C.A.d.B.; Stern, J.M. Evidence and credibility: Full Bayesian significance test for precise hypotheses. *Entropy* **1999**, *1*, 99–110. [CrossRef]
15. Pereira, C.A.d.B.; Stern, J.M.; Wechsler, S. Can a Significance Test be genuinely Bayesian? *Bayesian Anal.* **2008**, *3*, 79–100. [CrossRef]
16. Pereira, C.A.d.B.; Stern, J.M. The e-value: A fully Bayesian significance measure for precise statistical hypotheses and its research program. *São Paulo J. Math. Sci.* **2020**, 1–19. [CrossRef]
17. Kelter, R. Simulation data for the analysis of Bayesian posterior significance and effect size indices for the two-sample t-test to support reproducible medical research. *BMC Res. Notes* **2020**, *13*, 1–7. [CrossRef]
18. Kelter, R. fbst: An R package for the Full Bayesian Significance Test for testing a sharp null hypothesis against its alternative via the e-value. *Behav. Res. Methods* **2021**, in press. doi:10.3758/s13428-021-01613-6 [CrossRef]
19. Madruga, M.R.; Esteves, L.G.; Wechsler, S. On the Bayesianity of Pereira-Stern tests. *Test* **2001**, *10*, 291–299. [CrossRef]
20. Diniz, M.; Pereira, C.A.B.; Polpo, A.; Stern, J.M.; Wechsler, S. Relationship between Bayesian and frequentist significance indices. *Int. J. Uncertain. Quantif.* **2012**, *2*, 161–172. [CrossRef]
21. Ly, A.; Wagenmakers, E.J. A Critical Evaluation of the FBST ev for Bayesian Hypothesis Testing. *Comput. Brain Behav.* **2021**, 1–8. [CrossRef]
22. Kelter, R. On the Measure-Theoretic Premises of Bayes Factor and Full Bayesian Significance Tests: A Critical Reevaluation. *Comput. Brain Behav.* **2021**, 1–11. [CrossRef]
23. Kleijn, B. *The Frequentist Theory of Bayesian Statistics*; Springer: Amsterdam, The Netherlands, 2020.
24. Rosenman, R.H.; Brand, R.J.; Jenkins, D.; Friedman, M.; Straus, R.; Wurm, M. Coronary heart disease in Western Collaborative Group Study. Final follow-up experience of 8 1/2 years. *JAMA* **1975**, *233*, 872–877. [CrossRef]
25. Rouder, J.N.; Speckman, P.L.; Sun, D.; Morey, R.D.; Iverson, G. Bayesian t tests for accepting and rejecting the null hypothesis. *Psychon. Bull. Rev.* **2009**, *16*, 225–237. [CrossRef]
26. Cohen, J. *Statistical Power Analysis for the Behavioral Sciences*, 2nd ed.; Routledge: Hillsdale, NJ, USA, 1988.
27. Schervish, M.J. *Theory of Statistics*; Springer Verlag: New York, NY, USA, 1995.
28. Bauer, H. *Measure and Integration Theory*; De Gruyter: Berlin, Germany, 2001.
29. Robert, C.P. *The Bayesian Choice*, 2nd ed.; Springer New York: Paris, France, 2007. [CrossRef]
30. Jeffreys, H. *Theory of Probability*, 1st ed.; The Clarendon Press: Oxford, UK, 1939.
31. Haldane, J.B.S. A note on inverse probability. *Math. Proc. Camb. Philos. Soc.* **1932**, *28*, 55–61. [CrossRef]
32. Popper, K. *The Logic of Scientific Discovery*; Routledge: London, UK; New York, NY, USA, 1959. [CrossRef]
33. Berger, J. *Statistical Decision Theory and Bayesian Analysis*; Springer: New York, NY, USA, 1985.
34. Rousseau, J. Approximating Interval hypothesis: p-values and Bayes factors. In *Bayesian Statistics*; Bernado, J., Berger, J., Dawid, A., Smith, A., Eds.; Oxford University Press: Valencia, Spain, 2007; Volume 8, pp. 417–452.
35. Rao, C.R.; Lovric, M.M. Testing point null hypothesis of a normal mean and the truth: 21st Century perspective. *J. Mod. Appl. Stat. Methods* **2016**, *15*, 2–21. [CrossRef]
36. Kelter, R. Bayesian and frequentist testing for differences between two groups with parametric and nonparametric two-sample tests. *Wires Comput. Stat.* **2020**, *13*, e1523. [CrossRef]
37. Good, I. Surprise index. In *Encyclopedia of Statistical Sciences*; Kotz, S., Johnson, N., Reid, C., Eds.; John Wiley & Sons: New York, NY, USA, 1988; Volume 7.
38. Good, I. C332. Surprise indexes and p-values. *J. Stat. Comput. Simul.* **1989**, *32*, 90–92. [CrossRef]
39. Good, I.J. C420. The existence of sharp null hypotheses. *J. Stat. Comput. Simul.* **1994**, *49*, 241–242. [CrossRef]
40. Stern, J.M. Significance tests, Belief Calculi, and Burden of Proof in legal and Scientific Discourse. *Front. Artif. Intell. Its Appl.* **2003**, *101*, 139–147.

physical sciences forum

Proceeding Paper

Regularization of the Gravity Field Inversion Process with High-Dimensional Vector Autoregressive Models [†]

Andreas Kvas * and Torsten Mayer-Gürr

Institute of Geodesy, Graz University of Technology, Steyrergasse 30, 8010 Graz, Austria; mayer-guerr@tugraz.at

* Correspondence: kvas@tugraz.at

† Presented at the 40th International Workshop on Bayesian Inference and Maximum Entropy Methods in Science and Engineering, online, 4–9 July 2021.

Abstract: Earth's gravitational field provides invaluable insights into the changing nature of our planet. It reflects mass change caused by geophysical processes like continental hydrology, changes in the cryosphere or mass flux in the ocean. Satellite missions such as the NASA/DLR operated Gravity Recovery and Climate Experiment (GRACE), and its successor GRACE Follow-On (GRACE-FO) continuously monitor these temporal variations of the gravitational attraction. In contrast to other satellite remote sensing datasets, gravity field recovery is based on geophysical inversion which requires a global, homogeneous data coverage. GRACE and GRACE-FO typically reach this global coverage after about 30 days, so short-lived events such as floods, which occur on time frames from hours to weeks, require additional information to be properly resolved. In this contribution we treat Earth's gravitational field as a stationary random process and model its spatio-temporal correlations in the form of a vector autoregressive (VAR) model. The satellite measurements are combined with this prior information in a Kalman smoother framework to regularize the inversion process, which allows us to estimate daily, global gravity field snapshots. To derive the prior, we analyze geophysical model output which reflects the expected signal content and temporal evolution of the estimated gravity field solutions. The main challenges here are the high dimensionality of the process, with a state vector size in the order of 10^3 to 10^4, and the limited amount of model output from which to estimate such a high-dimensional VAR model. We introduce geophysically motivated constraints in the VAR model estimation process to ensure a positive-definite covariance function.

Keywords: GRACE/GRACE-FO; gravity field recovery; vector autoregressive models

Citation: Kvas, A.; Mayer-Gürr, T. Regularization of the Gravity Field Inversion Process with High-Dimensional Vector Autoregressive Models. *Phys. Sci. Forum* **2021**, *3*, 7. https://doi.org/10.3390/psf2021003007

Academic Editors: Wolfgang von der Linden and Sascha Ranftl

Published: 7 December 2021

Publisher's Note: MDPI stays neutral with regard to jurisdictional claims in published maps and institutional affiliations.

1. Introduction

Earth's gravitational field is key quantity for observing the state and change of our planet. Its variations in time can be related to surface mass changes [1] and thus it provides insight into geophysical and climate relevant processes, for example, sea level rise [2], ice mass loss [3], or the terrestrial water cycle [4]. Since 2002, the dedicated satellite mission Gravity Recovery And Climate Experiment (GRACE) [5,6] and its successor GRACE Follow-On (GRACE-FO) [7] have monitored temporal variations of Earth's gravitational field and have provided an invaluable data record for climate and Earth system sciences. The standard data products of both missions are unconstrained and constrained monthly snapshots of potential or surface mass changes. The time period of one month is not chosen arbitrarily but is a consequence of the orbit geometry of the satellites. As the satellites do not directly observe the changes in gravitational attraction but rather provide very precise measurements of their absolute and relative motion, the underlying potential field must be determined by an inversion process. This inversion process requires a global, homogeneous data coverage which for GRACE and GRACE-FO is reached after about 30 days.

However, in recent years different approaches to derive sub-monthly gravity field variations which combine satellite measurements with prior information have been developed.

41

Notable examples are the Kalman smoother approach by [8], a moving window approach by [9], constrained surface mass estimates from the Center of Space Research (CSR) [10] and a Kalman filter approach for regional applications [11]. Data sets dervied with these approaches have seen various applications in Earth system sciences, for example, analysis of large scale flood events [12], evaluation of geophysical models [13,14] or near real-time flood monitoring [15].

In this contribution we show how maximum entropy spectral estimation can be used to derive prior information for regularizing the inversion process for daily gravity field variations. To that end, we estimate a vector autoregressive (VAR) model from a time series of geophysical model output which approximates the expected signal. Following [15], we show how this spatio-temporal covariance information can be introduced into the inversion process and present a time series of daily gravity field variations derived from approximately thee years of GRACE-FO data.

2. Materials and Methods

A widely used method to recover Earth's gravitational field from a set of measurements is to solve an overdetermined system of equations in a least-squares adjustment [16]. Here, the measurements or observations $\mathbf{l}$ are related to unknown gravity field parameters through a (possibly non-linear) functional model $\mathbf{f}$ with:

$$\mathbf{l} = \mathbf{f}(\mathbf{x}) + \mathbf{e}. \tag{1}$$

We added the residual vector $\mathbf{e}$ because (1) equation is generally not consistent due to the stochastic nature of $\mathbf{l}$. To solve for $\mathbf{x}$, we expand $\mathbf{f}$ into a Taylor series around $\mathbf{x}_0$ and truncate it after the linear term. This yields the reduced observations equations:

$$\Delta\mathbf{l} = \mathbf{l} - \mathbf{f}(\mathbf{x}_0) = \mathbf{A}(\mathbf{x} - \mathbf{x}_0) + \mathbf{e}, \quad \mathbf{e} \sim \mathcal{N}(\mathbf{0}, \mathbf{\Sigma}_{\Delta\mathbf{l}}), \tag{2}$$

where the matrix $\mathbf{A}$ consists of the partial derivatives of $\mathbf{f}$ with respect to the parameter vector $\mathbf{x}$. In (2), the residual vector $\mathbf{e}$ is assumed to be centered and normally distributed with the covariance matrix $\mathbf{\Sigma}_{\Delta\mathbf{l}}$. This system of linear equations can be solved for the parameter corrections $\Delta\hat{\mathbf{x}} = \hat{\mathbf{x}} - \mathbf{x}_0$ by forming the system of normal equations:

$$(\mathbf{A}^T \mathbf{\Sigma}_{\Delta\mathbf{l}}^{-1} \mathbf{A})\Delta\hat{\mathbf{x}} = \mathbf{A}^T \mathbf{\Sigma}_{\Delta\mathbf{l}}^{-1} \Delta\mathbf{l}, \tag{3}$$

or more concise, $\mathbf{N}\Delta\hat{\mathbf{x}} = \mathbf{n}$. For this study, we express the changes Earth's gravitational field as spherical harmonic series expansion:

$$\Delta V(r, \vartheta, \lambda) = \frac{GM}{R} \sum_{n=0}^{\infty} \sum_{m=-n}^{n} \left(\frac{R}{r}\right)^{n+1} a_{nm} Y_{nm}(\vartheta, \lambda), \tag{4}$$

where the fully normalized surface spherical harmonics Y_{nm} are given by:

$$Y_{nm} = \begin{cases} P_{nm}(\cos\vartheta)\cos\lambda & \text{if } m \geq 0 \\ P_{n|m|}(\cos\vartheta)\sin\lambda & \text{if } m < 0 \end{cases}, \tag{5}$$

with P_{nm} being the fully normalized associated Legendre functions [17]. In practice, the series is truncated at a maximum degree $n_{\max}$, which depends on the application and is chosen here with $n_{\max} = 40$. Furthermore, we set degrees $n = \{0, 1\}$ to zero because we assume that Earth's mass does not vary and we fix the coordinate reference frame in Earth's center of mass. The unknown parameter corrections $\Delta\mathbf{x}$ therefore consist of the spherical harmonic coefficients a_{nm} with $n \in \{2, \ldots, 40\}$.

Since we want to determine daily temporal changes in Earth's gravitational field we set up this least-squares adjustment for a time series $t_i = i\Delta t, i \in \{0, \ldots, N-1\}$, where Δt is 1 day. We denote the resulting observation equations for time step i with a corresponding subscript: $\Delta l_i = A_i \Delta x_i + e_i$. All epochs can be combined in a single least squares adjustment with the block-diagonal normal equation coefficient matrix $\bar{N}_{ii} = N_i$ and the right hand side $\bar{n} = n_i$. Given the block-diagonal structure of $\bar{N}$ we can solve for each time step individually resulting in the estimated parameter corrections $\Delta \hat{x}_i$ and their corresponding variance-covariance matrix $\hat{\Sigma}_{\Delta \hat{x}_i} = N_i^{-1}$. Next to the input data quality, the uncertainty of $\Delta \hat{x}_i$ primarily depends on the spatial data coverage within the observation interval. To recover the full spherical harmonic spectrum, a global homogeneous data coverage is required. For the application here, where the observation time span is a single day, the satellites perform approximately 15 revolutions and thus we can only infer spherical harmonic coefficients up to order $m = 15$ [18]. The effect this sparse data coverage has on the estimated gravity field solution can be seen in Figure 1, where we propagated the uncertainty of an unconstrained daily GRACE-FO solution to equivalent water height (EWH) in space domain.

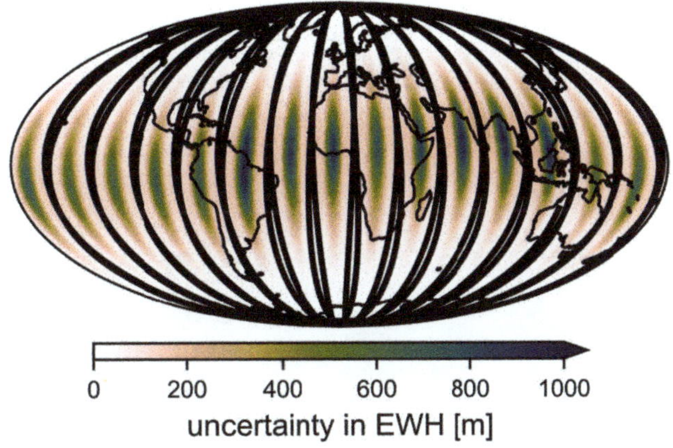

Figure 1. Uncertainty of an unconstrained daily gravity field estimated from one day of GRACE-FO data propagated to equivalent water height (EWH) in space domain. The solid black lines show the satellite ground tracks for this given day.

The uncertainty between the satellite ground tracks is magnitudes higher than the expected signal, which is in the order of ± 25 cm. Consequently, to recovery global, sub-monthly gravity field variations, additional external information is required.

We introduce prior information on the parameter corrections $\Delta x = [\Delta x_0^T, \ldots, \Delta x_{N-1}^T]^T$ in the form of a positive definite covariance matrix R. We further assume that Δx_i is a Gaussian, temporally stationary random process with an expected value of zero. This implies that their temporal covariance function $\Sigma_{\Delta x}(t_i, t_j)$ only depends on the lag h between the epoch i and j [19]. The prior distribution is then given by:

$$\Delta x \sim \mathcal{N}(0, R). \tag{6}$$

If we sort all epochs in temporally ascending order, the variance-covariance matrix R of the resulting vector is block-Toeplitz with the individual blocks given by

$$R_{ij} = \begin{cases} \Sigma_{\Delta x}(h = |j - i|) \text{ if } i \leq j \\ \Sigma_{\Delta x}^T(h = |j - i|) \text{ else} \end{cases}. \tag{7}$$

This prior information is then incorporated into the least squares adjustment as pseudo-observations of the form:

$$0 = \Delta x + v \quad v \sim \mathcal{N}(0, R), \tag{8}$$

with $\mathbf{v} = [\mathbf{v}_0^T, \ldots, \mathbf{v}_{N-1}^T]^T$. The constrained system of normal equations for the whole time series is subsequently given by:

$$(\bar{\mathbf{N}} + \mathbf{R}^{-1})\Delta\hat{x} = \bar{\mathbf{n}}. \tag{9}$$

The solution to this augmented least-squares problem is numerically equivalent to the Bayes estimate under the assumption of the given prior information.

Both GRACE and GRACE-FO exhibit an approximate repeat cycle of 4–5 days at the equator. As the observation information is closely tied to the ground track coverage es presented in Figure 1, this means that points at the equator are also updated at this repeat frequency. Between these updates, the solutions are supported by the introduced temporal constraint and slowly tend towards the Taylor series expansion point $\mathbf{x}_0$. We chose $\mathbf{x}_0$ as a long-term estimate of the secular and seasonal gravity field variations, which captures a large part of the total signal. In higher latitudes, where the ground tracks converge, observation updates occur more frequently. This means that as we move closer to the poles, the gravity field solutions become less reliant on the prior information.

Note that the covariance function $\Sigma_{\Delta\mathbf{x}}(h)$ is still unknown at this point. To estimate the stochastic properties we perform maximum entropy spectral estimation by fitting a vector autoregressive model (VAR) with:

$$\Delta\mathbf{x}_i = \sum_{k=1}^{p} \boldsymbol{\Phi}_k \Delta\mathbf{x}_{i-k} + \mathbf{w}_i, \qquad \mathbf{w}_i \sim \mathcal{N}0, \Sigma_\mathbf{w} \tag{10}$$

to a time series of geophysical model output $\mathbf{m}_i$, which approximates the true spatio-temporal covariance structure of Earth's time-variable gravity field. We make use of the Earth System Model of the European Space Agency (ESA ESM, [20]) to generate daily spherical harmonics coefficients vectors. To estimate the VAR model, we use Yule–Walker equations [19] to estimate the empirical covariance matrices for lags $h \in \{0, \ldots, 3\}$ using the unbiased estimator:

$$\hat{\Sigma}(h) = \frac{1}{M-h} \sum_{k=0}^{M-h-1} \mathbf{m}_k \mathbf{m}_{k+h}^T. \tag{11}$$

After estimating the empirical covariance function and before solving the Yule–Walker equations, we apply geophysically motivated constraints onto the obtained covariance matrices. Specifically, we want set correlations between land and ocean to zero and reduce the overall correlation length. These constraints are very easy to introduce in space domain so we first propagate $\hat{\Sigma}(h)$ from the spherical harmonics domain to an EWH grid in space domain. Now the matrices can be decomposed into $\sigma_{ij}(h) = r_{ij}(h)\sigma_i\sigma_j$, where $r_{ij}(h)$ represents the correlation between two grid points at lag h and $\sigma_{i/j}$ is the standard deviation of the corresponding grid points. The constraints are then realized by setting $r_{ij}(h) = 0$ if the points i and j are on different domains and applying a decay function $\tilde{r}_{ij}(h) = r_{ij}(h)e^{-\psi/\psi_0}$. The amount at which the correlation between points in dependence of their spherical distance ψ is reduced is governed by the the parameter ψ_0 which we chose as 1100 km. After propagating the modified empirical covariance matrices back to spherical harmonics domain, we can now solve the Yule–Walker equations to obtain the VAR model coefficients $\boldsymbol{\Phi}_k$ and compute the white noise covariance matrix $\Sigma_\mathbf{w}$. Now we could compute the covariance function for all lags h, however, given the high dimensionality of the problem, assembling the full block variance-covariance matrix is not feasible. Instead we use the VAR model to transform the pseudo-observation equations (8). For each time step i, we apply $0 = -\Delta\mathbf{x}_i + \sum_{k=1}^{p} \boldsymbol{\Phi}_k\Delta\mathbf{x}_{i-k}$, which can also be written as a lower triangular block matrix applied to the full time series $\Delta\mathbf{x}$ as in:

$$0 = \bar{\mathbf{\Phi}}\Delta\mathbf{x} + \mathbf{w} \quad \mathbf{w} \sim \mathcal{N}(0, \bar{\mathbf{\Sigma}}_{\mathbf{w}}), \tag{12}$$

where $\bar{\mathbf{\Sigma}}_{\mathbf{w}} = \bar{\mathbf{\Phi}}\mathbf{R}\bar{\mathbf{\Phi}}^{T}$ is a block diagonal matrix with $\mathbf{\Sigma}_{\mathbf{w}}$ on the main diagonal. This has the great advantage that the resulting normal equation coefficient matrix is block-banded with a bandwith of $p + 1$ and thus greatly reduces the memory demands. The transformed system of normal equations which we use to estimate the full time series of gravity field variations is then given by:

$$(\bar{\mathbf{N}} + \bar{\mathbf{\Phi}}^{T}\bar{\mathbf{\Sigma}}_{\mathbf{w}}^{-1}\bar{\mathbf{\Phi}})\Delta\hat{\mathbf{x}} = \bar{\mathbf{n}}. \tag{13}$$

Note that for a VAR model of order 1, (13) yields the same results as the Kalman smoother approach presented in [8].

3. Results

To showcase how the VAR regularization affects the gravity field estimates, we computed systems of normal equations from close to three years of GRACE-FO sensor data [21]. Figure 2 shows the autocovariance matrix computed from the VAR model and the temporal variability of the computed time series propagated to space domain.

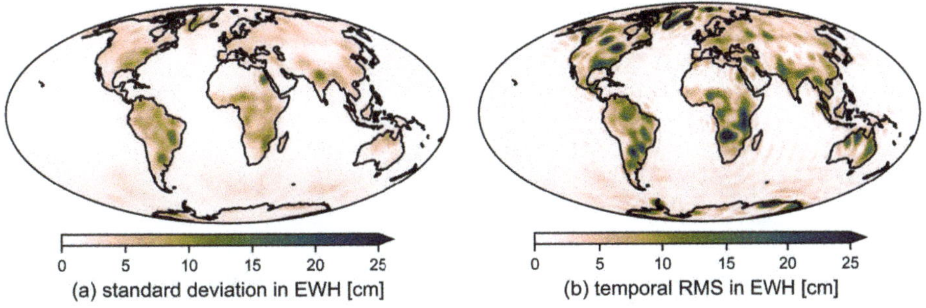

Figure 2. Comparison of (**a**) the autocovariance matrix derived from the estimated VAR model expressed as standard deviation per grid point and (**b**) the temporal RMS of the computed gravity field solutions.

As can be seen, the estimated gravity field time series shows a lot more signatures than the expected signal covariance. This means that the GRACE-FO observations provide significant information and are not overly constrained by the derived VAR model. It should be noted that the VAR model and subsequently the spatio-temporal constraints depend on the geophysical models used in their derivation. However, the influence of different hydrological models is very small as shown in [12]. This suggests that the daily gravity field solutions are primarily data driven.

To gauge how the daily gravity field variations compare with standard GRACE/GRACE-FO products, we compute area mean time series for selected hydrological basins. As monthly data set we use the unconstrained monthly solutions of ITSG-Grace_operational [22,23] computed at Graz University of Technology, filtered with DDK4 spatial filter [24].

Figure 3 shows the area mean time series of 4 regions where we see larger water storage variations. We can see that the daily gravity field solutions clearly pick up sub-monthly signal and thus provide additional information compared to the standard GRACE/GRACE-FO products.

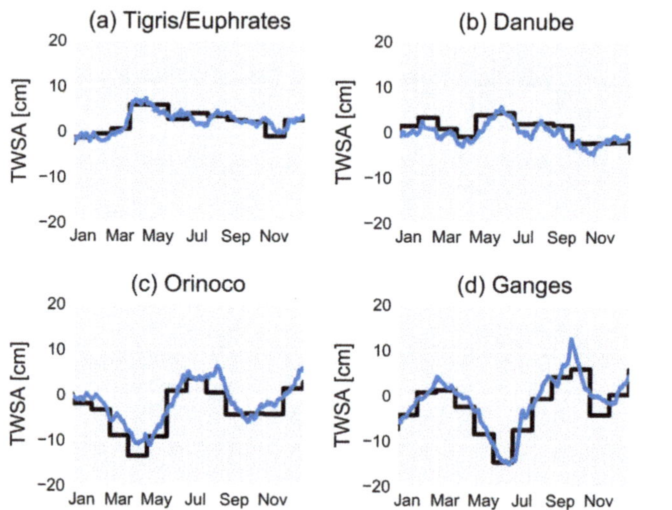

Figure 3. Total water storage anomaly (TWSA) in equivalent water height for (**a**) Tigris/Euhprates, (**b**) Danube, (**c**) Orinoco, and (**d**) Ganges basin for the year 2019 (trend and annual signal are removed). The estimated daily solutions are compared with DDK4-filtered monthly solutions from ITSG-Grace_operational shown as step function.

4. Discussion

We have shown that VAR models provide an efficient way of introducing spatio-temporal prior information into the gravity field inversion process. This approach is equivalent to constraining the gravity field estimates with the full block-Toeplitz variance-covariance matrix, however requires only a fraction of the memory. Constraining the inversion of daily gravity field variations with a VAR model of arbitrary order p constitutes the generalization of the Kalman smoother approach of [8], which is the special case for $p = 1$. If the VAR model is derived as a maximum entropy spectral estimate from a time series of geophysical model output, a few challenges arise. The high-dimensionality of the problem with a state vector size in the order of 10^3 to 10^4, combined with the limited availability of model output results in a very low redundancy in the VAR model estimation. We counteract this by introducing geophysically motivated constraints on the empirical covariance matrix estimates. First, we decouple the land and ocean domain and then we reduce the remaining correlation between far away point by introducing an exponential decay function dependent on the spherical distance. This drastically improves the condition of the VAR estimation problem. Finally, we show that the derived prior information in combination with GRACE-FO measurement data yields reasonable results by computing a time series of daily gravity field variations from June 2018 to March 2021. Here, we can clearly see sub-monthly signals which underlines the proficiency of the approach.

Author Contributions: A.K. worked on the theoretical background, performed the gravity field recovery and wrote the manuscript; T.M.-G. worked on the theoretical background. All authors have read and agreed to the published version of the manuscript.

Data Availability Statement: The full time series of gravity field variations and the estimated autoregressive model are available on ftp://ftp.tugraz.at/outgoing/ITSG/GRACE/ITSG-Grace_operational/daily_kalman (accessed on 21 March 2021).

Conflicts of Interest: The authors declare no conflict of interest.

Abbreviations

The following abbreviations are used in this manuscript:

GRACE	Gravity Recovery And Climate Experiment
GRACE-FO	GRACE Follow-On
CSR	Center for Space Research
VAR	vector autoregressive
EWH	equivalent water height
RMS	root mean square

References

1. Wahr, J.; Molenaar, M.; Bryan, F. Time variability of the Earth's gravity field: Hydrological and oceanic effects and their possible detection using GRACE. *J. Geophys. Res. Solid Earth* **1998**, *103*, 30205–30229. [CrossRef]
2. Dobslaw, H.; Dill, R.; Bagge, M.; Klemann, V.; Boergens, E.; Thomas, M.; Dahle, C.; Flechtner, F. Gravitationally Consistent Mean Barystatic Sea Level Rise From Leakage-Corrected Monthly GRACE Data. *J. Geophys. Res. Solid Earth* **2020**, *125*, e2020JB020923. [CrossRef]
3. Groh, A.; Horwath, M.; Horvath, A.; Meister, R.; Sørensen, L.S.; Barletta, V.R.; Forsberg, R.; Wouters, B.; Ditmar, P.; Ran, J.; et al. Evaluating GRACE Mass Change Time Series for the Antarctic and Greenland Ice Sheet—Methods and Results. *Geosciences* **2019**, *9*, 415. [CrossRef]
4. Eicker, A.; Forootan, E.; Springer, A.; Longuevergne, L.; Kusche, J. Does GRACE see the terrestrial water cycle "intensifying"? *J. Geophys. Res. Atmos.* **2016**, *121*, 733–745. [CrossRef]
5. Tapley, B.D.; Bettadpur, S.; Watkins, M.; Reigber, C. The gravity recovery and climate experiment: Mission overview and early results. *Geophys. Res. Lett.* **2004**, *31*. [CrossRef]
6. Tapley, B.D.; Watkins, M.M.; Flechtner, F.; Reigber, C.; Bettadpur, S.; Rodell, M.; Sasgen, I.; Famiglietti, J.S.; Landerer, F.W.; Chambers, D.P.; et al. Contributions of GRACE to understanding climate change. *Nat. Clim. Chang.* **2019**, *9*, 358–369. [CrossRef] [PubMed]
7. Landerer, F.W.; Flechtner, F.M.; Save, H.; Webb, F.H.; Bandikova, T.; Bertiger, W.I.; Bettadpur, S.V.; Byun, S.H.; Dahle, C.; Dobslaw, H.; et al. Extending the Global Mass Change Data Record: GRACE Follow-On Instrument and Science Data Performance. *Geophys. Res. Lett.* **2020**, *47*, e2020GL088306. [CrossRef]
8. Kurtenbach, E.; Eicker, A.; Mayer-Gürr, T.; Holschneider, M.; Hayn, M.; Fuhrmann, M.; Kusche, J. Improved daily GRACE gravity field solutions using a Kalman smoother. *J. Geodyn.* **2012**, *59–60*, 39–48. [CrossRef]
9. Sakumura, C.; Bettadpur, S.; Save, H.; McCullough, C. High-frequency terrestrial water storage signal capture via a regularized sliding window mascon product from GRACE. *J. Geophys. Res. Solid Earth* **2016**, *121*, 4014–4030. [CrossRef]
10. Save, H. *CSR RL05 GRACE Daily Swath Mass Anomaly Estimates over the Ocean*; Center for Space Research, The University of Texas at Austin: Austin, TX, USA, 2019. [CrossRef]
11. Ramillien, G.; Seoane, L.; Schumacher, M.; Forootan, E.; Frappart, F.; Darrozes, J. Recovery of Rapid Water Mass Changes (RWMC) by Kalman Filtering of GRACE Observations. *Remote Sens.* **2020**, *12*, 1299. [CrossRef]
12. Gouweleeuw, B.; Kvas, A.; Gruber, C.; Gain, A.; Mayer-Gürr, T.; Flechtner, F.; Güntner, A. Daily GRACE gravity field solutions track major flood events in the Ganges-Brahmaputra Delta. *Hydrol. Earth Syst. Sci.* **2018**, *22*, 2867–2880. [CrossRef]
13. Eicker, A.; Jensen, L.; Wöhnke, V.; Dobslaw, H.; Kvas, A.; Mayer-Gürr, T.; Dill, R. Daily GRACE satellite data evaluate short-term hydro-meteorological fluxes from global atmospheric reanalyses. *Sci. Rep.* **2020**, *10*, 4504. [CrossRef] [PubMed]
14. Schindelegger, M.; Harker, A.A.; Ponte, R.M.; Dobslaw, H.; Salstein, D.A. Convergence of Daily GRACE Solutions and Models of Submonthly Ocean Bottom Pressure Variability. *J. Geophys. Res. Oceans* **2021**, *126*, e2020JC017031. [CrossRef]
15. Kvas, A. Estimation of High-Frequency Mass Variations from Satellite Data in near Real-Time: Implementation of a Technology Demonstrator for near Real-Time GRACE/GRACE-FO Gravity Field Solutions. Ph.D. Thesis, Graz University of Technology, Graz, Austria, 2020. [CrossRef]
16. Koch, K.R. Least squares adjustment and collocation. *Bulletin Géodésique* **1977**, *51*, 127–135. [CrossRef]
17. Heiskanen, W.A.; Moritz, H. Physical geodesy. *Bulletin Géodésique (1946–1975)* **1967**, *86*, 491–492. [CrossRef]
18. Weigelt, M.; Sneeuw, N.; Schrama, E.J.; Visser, P.N. An improved sampling rule for mapping geopotential functions of a planet from a near polar orbit. *J. Geod.* **2013**, *87*, 127–142. [CrossRef]
19. Lütkepohl, H. *New Introduction to Multiple Time Series Analysis*; Springer: Berlin/Heidelberg, Germany, 2005; Volume 1, pp. 1–764.
20. Dobslaw, H.; Bergmann-Wolf, I.; Dill, R.; Forootan, E.; Klemann, V.; Kusche, J.; Sasgen, I. The updated ESA Earth System Model for future gravity mission simulation studies. *J. Geod.* **2015**, *89*, 505–513. [CrossRef]
21. NASA Jet Propulsion Laboratory. GRACE-FO Level-1B Release Version 4.0 from JPL in ASCII. 2019. Available online: https://podaac-tools.jpl.nasa.gov/drive/files/allData/gracefo/L1B/JPL/RL04/ASCII (accessed on 21 March 2021).

22. Kvas, A.; Behzadpour, S.; Ellmer, M.; Klinger, B.; Strasser, S.; Zehentner, N.; Mayer-Gürr, T. ITSG-Grace2018: Overview and Evaluation of a New GRACE-Only Gravity Field Time Series. *J. Geophys. Res. Solid Earth* **2019**, *124*, 9332–9344. [CrossRef]
23. Mayer-Gürr, T.; Behzadpur, S.; Ellmer, M.; Kvas, A.; Klinger, B.; Strasser, S.; Zehentner, N. *ITSG-Grace2018—Monthly, Daily and Static Gravity Field Solutions from GRACE*; GFZ Data Services: Potsdam, Germany, 2018. [CrossRef]
24. Kusche, J.; Schmidt, R.; Petrovic, S.; Rietbroek, R. Decorrelated GRACE time-variable gravity solutions by GFZ, and their validation using a hydrological model. *J. Geod.* **2009**, *83*, 903–913. [CrossRef]

physical sciences forum

Proceeding Paper

Survey Optimization via the Haphazard Intentional Sampling Method [†]

Miguel Miguel [1], Rafael Waissman [2], Marcelo Lauretto [2] and Julio Stern [1,*]

[1] Institute of Mathematics and Statistics, University of São Paulo, São Paulo 05508-900, Brazil; mgabriel@ime.usp.br

[2] School of Arts, Sciences and Humanities, University of São Paulo, São Paulo 03828-000, Brazil; rafaelwaissman@usp.br (R.W.); marcelolauretto@usp.br (M.L.)

* Correspondence: jstern@ime.usp.br

† Presented at the 40th International Workshop on Bayesian Inference and Maximum Entropy Methods in Science and Engineering, online, 4–9 July 2021.

Abstract: In previously published articles, our research group has developed the *Haphazard Intentional Sampling* method and compared it to the *Rerandomization* method proposed by K.Morgan and D.Rubin. In this article, we compare both methods to the pure randomization method used for the Epicovid19 survey, conducted to estimate SARS-CoV-2 prevalence in 133 Brazilian Municipalities. We show that Haphazard intentional sampling can either substantially reduce operating costs to achieve the same estimation errors or, the other way around, substantially improve estimation precision using the same sample sizes.

Keywords: haphazard intentional sampling; rerandomization; pure randomization; optimal sampling design

Citation: Miguel, M.; Waissman, R.; Lauretto, M.; Stern, J. Survey Optimization via the Haphazard Intentional Sampling Method. *Phys. Sci. Forum* **2021**, *3*, 4. https://doi.org/10.3390/psf2021003004

Academic Editors: Wolfgang von der Linden and Sascha Ranftl

Published: 5 November 2021

Publisher's Note: MDPI stays neutral with regard to jurisdictional claims in published maps and institutional affiliations.

1. Introduction

Traditional sampling with face-to-face interviews usually demands large staff and infrastructure and expensive field operations to cover a representative group of the population of interest. Even then, pure (or stratified) randomized experiments do not guarantee efficient control over specific sets of covariates, and there may be large divergences between sample and population statistics. To address this problem, Lauretto et al. [1,2] and Fossaluza et al. [3] developed the *Haphazard Intentional Sampling* method, an approach that combines intentional sampling, using methods of numerical optimization for an appropriate objective function, with random disturbances ensuring good decoupling properties. The word *haphazard* was used by Dennis Lindley to distinguish the decoupling effect from the tool used to obtain the desired decoupling, namely, randomization; for further details, see [1,3,4]. For a fixed sample size, this technique aims at diminishing the distance between sample and population regarding specific covariates of interest or, the other way around, minimizing the sample size needed to achieve good enough expected agreement between sample and population regarding specific covariates of interest. The Mahalanobis distance is the natural choice for the statistical model at hand, but other L_p distances, or convex combinations thereof, will be used as approximations useful for numerical computation, as explained in the following sections. This method can be applied in several contexts, such as allocations of treatment and control groups in medical trials [2] or in statistical sampling problems [5]. In this method, a weight factor, λ, adjusts the weight of the random perturbation relative to the deterministic objective function of the optimization problem. In practical problems, the weight factor λ can be calibrated in such a way that, on the one hand, it is small enough to generate only slightly sub-optimal solutions and, on the other hand, it is large enough to break potential confounding effects that could introduce spurious statistical biases in the study.

In this paper, the performance of the Haphazard intentional sampling method is compared to pure random sampling and to the *Rerandomization* methods proposed by Morgan and Rubin [6]. As a benchmark, we use an artificial sampling problem concerning inferences about the prevalence of Sars-CoV-2, using covariates from public data sets generated by the 2010 Census of the Brazilian Institute of Geography and Statistics (IBGE). Pending analyses of this benchmark study, real epidemiological applications shall be developed in the near future. Sampling procedures of the three aforementioned methods were repeatedly applied to the benchmark problem, in order to obtain performance statistics regarding how well generated samples represent the population.

2. Haphazard Intentional Sampling Method

In this section, we present the formulation of the Haphazard intentional sampling method presented by Lauretto et al. [1,2]. Let $\mathbf{X}$ denote a matrix in $\mathbb{R}^{n \times d}$, where n is the number of candidate sampling units and d is the number of covariates of interest. An allocation consists of assigning to each unit a group chosen from a set of possible groups, $\mathcal{G} = \{0, 1, 2, ...g\}$. We denote by $\mathbf{w}$ an allocation vector in $\mathcal{G}^n$, assigning each unit to a group. For simplicity, we assume only two groups, that is, $\mathcal{G} = \{0, 1\}$, the control and treatment groups, or the unsampled and sampled units. We also assume that the number of units assigned to each group is previously defined. That is, integers n_1 and n_0 exist such that $n_1 + n_0 = n$, $\mathbf{1}\mathbf{w}^t = n_1$ and $\mathbf{1}(1 - \mathbf{w})^t = n_0$. $\mathbf{1}$ denotes a vector of ones with the proper size; therefore, the scalar product $\mathbf{1}\mathbf{w}^t$ is the sum of the scalar components of $\mathbf{w}$. The goal of the allocation problem is to generate an allocation, $\mathbf{w}$, that, with high probability, approximately minimizes the imbalance between groups with respect to a loss function, $L(\mathbf{w}, \mathbf{X})$.

The Mahalanobis distance is the metric of choice for statistical models based on the multivariate normal distribution; for further details, see Stern [7] (Section 6.2). The Mahalanobis distance between the covariates of interest in each group is defined as follows. Let $\mathbf{A}$ be an arbitrary matrix in $\mathbb{R}^{n \times m}$. Furthermore, define $\mathbf{A}^* := \mathbf{AL}$, where $\mathbf{L}$ is the lower triangular Cholesky factor [8] of the inverse of covariance matrix of $\mathbf{A}$; that is, $\mathrm{Cov}(\mathbf{A})^{-1} = \mathbf{LL}^t$.

For an allocation $\mathbf{w}$, let $\overline{\mathbf{A}^*}^1$ and $\overline{\mathbf{A}^*}^0$ denote the averages of each column of $\mathbf{A}^*$ over units allocated to, respectively, groups 1 and 0 according to the row vector $\mathbf{w}$:

$$\overline{\mathbf{A}^*}^1 := (1/n_1)\mathbf{w}\mathbf{A}^* \quad \text{and} \quad \overline{\mathbf{A}^*}^0 := (1/n_0)(1 - \mathbf{w})\mathbf{A}^* . \tag{1}$$

The Mahalanobis distance between the average of the column values of $\mathbf{A}$ in each group specified by $\mathbf{w}$ is defined as:

$$M(\mathbf{w}, \mathbf{A}) := m^{-1}\|\overline{\mathbf{A}^*}^1 - \overline{\mathbf{A}^*}^0\|_2 , \tag{2}$$

where m denotes the number of columns of A.

2.1. Pure Intentional Sampling Formulation

Under the Mahalanobis loss function, a pure intentional sampling procedure consists of generating an allocation $\mathbf{w}$ that minimizes the following optimization problem:

$$\begin{aligned}
\text{minimize} \quad & M(\mathbf{w}, \mathbf{X}) \\
\text{subject to} \quad & \mathbf{1}\mathbf{w}^t = n_1 \\
& \mathbf{1}(1 - \mathbf{w})^t = n_0 \\
& \mathbf{w} \in \{0, 1\}^n
\end{aligned} \tag{3}$$

The formulation presented in Equation (3) is a Mixed-Integer Quadratic Programming Problem (MIQP) [9], that can be computationally very expensive. The *hybrid loss function*, $H(\mathbf{w}, \mathbf{A})$, is a surrogate function for $M(\mathbf{w}, \mathbf{A})$ built using a linear combination of L_1 and L_∞ norms; see Ward and Wendell [10]:

$$H(\mathbf{w}, \mathbf{A}) := m^{-1}\left(\|\overline{\mathbf{A}^*}^1 - \overline{\mathbf{A}^*}^0\|_1 + \sqrt{m}\ \|\overline{\mathbf{A}^*}^1 - \overline{\mathbf{A}^*}^0\|_\infty\right) \tag{4}$$

The minimization of $H(\mathbf{w}, \mathbf{A})$ yields the Mixed-Integer Linear Programming Problem (MILP) defined in the next equation, which is computationally much less expensive than the MIQP problem (3); see Murtagh [11] and Wolsey and Nemhauser [9].

$$\begin{aligned} \text{minimize} \quad & H(\mathbf{w}, \mathbf{X}) \\ \text{subject to} \quad & \mathbf{1}\mathbf{w}^t = n_1 \\ & \mathbf{1}(\mathbf{1} - \mathbf{w})^t = n_0 \\ & \mathbf{w} \in \{0, 1\}^n \end{aligned} \tag{5}$$

Statistical inference based on pure intentional sampling is vulnerable to malicious manipulation, unconscious biases, and many other confounding effects. In the Frequentist School of statistics, the use of intentional allocation is anathema, whereas in the Bayesian School, it has been the subject of long-standing debates. The solution presented in this paper is a compromise aiming to achieve the effective performance of intentional sampling but using moderate randomization to avoid systematic confounding effects. Lauretto et al. [1] and Fossaluza et al. [3] give a thorough discussion of the motivation and history of the ideas leading to the Haphazard intentional sampling method.

2.2. Haphazard Formulation

The Haphazard intentional sampling method consists of extending the pure intentional sampling method, formulated in Equation (5), as a MILP optimization problem, with the introduction of a noisy component. Let $\mathbf{Z}$ be an artificially generated random matrix in $\mathbb{R}^{n \times k}$, with elements that are independent and identically distributed according to the standard normal distribution. For a given tuning parameter, $\lambda \in [0, 1]$, the Haphazard method, aims to solve the following optimization problem:

$$\begin{aligned} \text{minimize} \quad & (1 - \lambda)\,H(\mathbf{w}, \mathbf{X}) + \lambda\,H(\mathbf{w}, \mathbf{Z}) \\ \text{subject to} \quad & \mathbf{1}\mathbf{w}^t = n_1 \\ & \mathbf{1}(\mathbf{1} - \mathbf{w})^t = n_0 \\ & \mathbf{w} \in \{0, 1\}^n \end{aligned} \tag{6}$$

The parameter λ controls the amount of perturbation that is added to the surrogate loss function, $H(\mathbf{w}, \mathbf{X})$. If $\lambda = 0$, then $\mathbf{w}^*$ is the deterministic optimal solution for $H(\mathbf{w}, \mathbf{X})$, corresponding to the pure intentional sampling. If $\lambda = 1$, then $\mathbf{w}^*$ is the optimal solution for the artificial random loss, $H(\mathbf{w}, \mathbf{Z})$, corresponding to a completely random allocation. By choosing an intermediate value of λ (as discussed in Section 3.2), one can obtain $\mathbf{w}^*$ to be a partially randomized allocation such that, with a high probability, $H(\mathbf{w}^*, \mathbf{X})$ is close to the minimum loss.

3. Case Study

The artificial data set used for the simulations carried in this study are inspired by Epicovid19 [12], a survey conducted by the Brazilian Institute of Public Opinion and Statistics (IBOPE) and the Federal University of Pelotas (UFPel) to estimate SARS-CoV-2 infection prevalence in 133 Brazilian municipalities. Our study is supplemented by data from the 2010 Brazilian census conducted by IBGE, giving socio-economic information by *census sector*. Sectors are the minimal units by which census information is made publicly available. Typically, each sector has around 200 households. Furthermore, households in a sector form a contiguous geographic area with approximately homogeneous characteristics.

The first step of the sampling procedure of Epicovid19 study consisted of randomly selecting a subset of census sectors of each surveyed municipality. As a second step, at each of the selected sectors, a subset of households was randomly selected for a detailed interview concerning socio-economic characteristics and SARS-CoV-2 antibody testing.

Our benchmark problem is based on the original Epicovid19 study, where we evaluate the impact of alternative census sector sampling procedures on the estimation of the response variable, namely, SARS-CoV-2 prevalence. In order to simulate outcomes for alternative sector selections, we used an auxiliary regression model for this response variable, as explained in the sequel.

3.1. Auxiliary Regression Model for SARS-CoV-2 Prevalence

The auxiliary regression model for SARS-CoV-2 prevalence had the Epicovid19 estimated infection rates adjusted for the spread of the pandemic in subsequent months and corrected for underreporting due to lack of intensive testing in Brazil. As explanatory variables, this auxiliary model used 15 socio-demographic covariates, including income, ethnicity, age, sanitation condition, etc. The parameters of this auxiliary regression model were estimated using standard regression packages available at the R statistical environment. Since the response variable is simulated by this auxiliary regression model, its covariates and their weight coefficients in the regression can be taken as a valid representativity target, that is, the Haphazard and Rerandomization methods will try to make sector selections that resemble the population characteristics corresponding to these 15 covariates.

The auxiliary model was a *logit* link regression, specified by selecting three of the most relevant predictive variables, namely, average income, population percentage with zero income, and percentage of households with two or more bathrooms (a standard indirect measure of wealth used by IBGE):

$$ln(p_i/(1-p_i)) = \eta_i = \hat{\beta}_0 + \hat{\beta}_1 x_{i,1} + \hat{\beta}_2 x_{i,2} + \hat{\beta}_3 x_{i,3} + \epsilon_i \tag{7}$$

$$p_i = \frac{e^{\eta_i}}{(1+e^{\eta_i})} \tag{8}$$

p_i: simulated SARS-CoV-2 prevalence in sector i;
$x_{i,1}$: income in census sector i;
$x_{i,2}$: zero income population percentage in census sector i;
$x_{i,3}$: percentage of households with two or more bathrooms in census sector i.

3.2. Balance and Decoupling Trade-Off in the Haphazard Method

The Haphazard intentional sampling method is not exclusively concerned with choosing maximally representative samples. Equally important is to prevent estimation biases induced by spurious confounding effects. This is exactly the role of the *decoupling* effects engendered by standard randomization procedures. We need a quantitative measure to assess how effectively the noise introduced in the method, with weight λ, is performing this task. A proxy measure of this sort can be constructed using Fleiss' Kappa coefficient, conceived to measure the degree of agreement between nominal scales assigned by multiple raters, see Fleiss [13]. In our context, it is used as follows.

For r repetitions of a sampling procedure, let $r_{i,j}$ denote the number of times element $i \in \{1, 2, \ldots, N\}$ is allocated to group $j \in \{0, 1\}$. Let $\overline{P}_o$ denote the observed average proportion of concordance among all allocation pairs. Let $\overline{P}_e$ denote the expected agreement that would be obtained by chance, conditional on the proportion of assignments that were observed in each group j.

$$\overline{P}_o = \frac{1}{Nr(r-1)} \sum_{i=1}^{N} \sum_{j=0}^{1} r_{i,j}(r_{i,j}-1) \qquad \overline{P}_e = \sum_{j=0}^{1} \frac{\left(\sum_{i=1}^{N} r_{i,j}\right)^2}{(Nr)^2} \tag{9}$$

The Fleiss' Kappa coefficient is obtained by the ratio of the difference between the observed and the expected random agreement, $\overline{P}_o - \overline{P}_e$, and the difference between total agreement and the agreement obtained by chance, $1 - \overline{P}_e$:

$$\kappa = \frac{\overline{P}_o - \overline{P}_e}{1 - \overline{P}_e} \tag{10}$$

The relation between decoupling and the degree of disturbance added is assessed empirically. The following transformation between parameters λ and λ^* is devised to equilibrate the weights given to the terms of Equation (6) corresponding to the covariates of interest and artificial noise, according to dimensions d (the number of columns of $\mathbf{X}$) and k (the number of columns of $\mathbf{Z}$).

$$\lambda = \lambda^* / \left[\lambda^*(1 - k/d) + k/d \right], \quad \text{where} \quad \lambda^* \in \{0.005, 0.01, 0.05, 0.1, 0.25, 0.5\}. \tag{11}$$

The trade-off between balancing and decoupling also varies according to the characteristics of each municipality. Small municipalities have only a limited number of census sectors and, hence, also a limited set of near-optimal solutions. Therefore, for small municipalities, good decoupling requires a larger λ^*. Figure 1a shows, for the smallest of the 133 municipalities in the database (with 34 census sectors), the trade-off between balance and decoupling (Fleiss' Kappa) as λ^* varies in proper range. Figure 1b shows the same trade-off for a medium sized municipality. Since it has many more sectors (176), it is a lot easier to find well-balanced solutions and, hence, good decoupling is a lot easier to achieve.

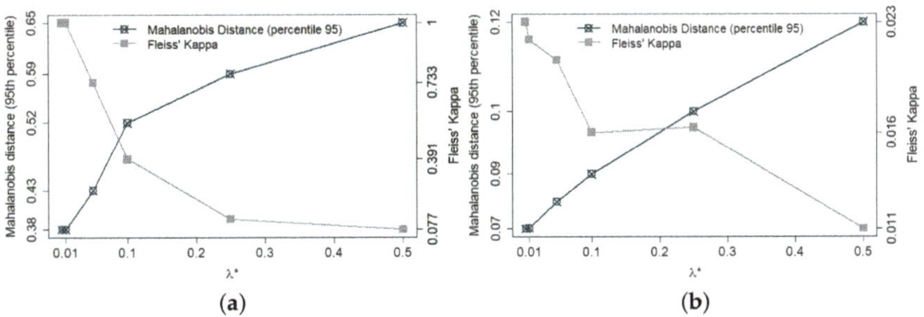

Figure 1. Trade-off between balance and decoupling in 300 allocations for two municipalities containing, respectively, 34 (panel **a**) and 176 (panel **b**) sectors. Sectors are the minimal units by which census information is made publicly available, each containing about 200 households. Balance between allocated and non-allocated sectors is expressed by the 95th percentile of Mahalanobis distance. Decoupling is expressed by Fleiss' Kappa coefficient—notice the different range in each case (**a**,**b**).

Larger municipalities engender larger optimization problems (for the number of binary decision variables equals the number of census sectors) that, in turn, usually require more CPU time for the MILP solver. Table 1 displays empirically calibrated parameters λ^* and maximum CPU times under the hardware configuration described in Section 3.3.

Table 1. Parameters λ^* and maximum CPU time for MILP solver by number of sectors.

Sectors	λ^*	Time (s)
<50	0.1	5
50–4000	0.01	30
>4000	0.001	120

3.3. Benchmark Experiments and Computational Setups

Our performance experiments used a subset of 10 municipalities of the 133 in the original Epicovid19 study, covering a wide range of population size and characteristics. Following the original Epicovid19 protocol, a sample of 25 census sectors was selected at each municipality. The sampling procedure for selecting these 25 sectors was repeated 300 times, using each of the three methods under comparison, namely, Haphazard method, Rerandomization, and pure randomization.

Numerical optimization and statistical computing tasks were implemented using the R v.3.6.1. environment [14] and the Gurobi v.9.0.1 optimization solvers [15]. The computer used to run these routines had an AMD RYZEN 1920X processor (3.5 GHz, 12 cores, 24 threads), ASROCK x399 motherboard, 64 GB DDR4 RAM, and Linux Ubuntu 18.04.5 LTS operating system. There is nothing particular about hardware configuration, with performance being roughly proportional to general computing power.

4. Experimental Results

In this section, we present the comparative results for the Haphazard, Rerandomization, and simple randomization methods, considering the metrics discussed in the sequel.

4.1. Group Unbalance among Covariates

We compute the standardized difference between group means for each covariate, based on 300 simulated allocations per method. Specifically, we compute the empirical distribution of the statistics $(\overline{X}_{\cdot j}^{1} - \overline{X}_{\cdot j}^{0})/s_j$, where $\overline{X}_{\cdot j}^{1}$ and $\overline{X}_{\cdot j}^{0}$ denote the averages of the j-th column of X over units allocated to, respectively, groups 1 and 0 (see Equation (1)); and s_j is the reference scale given by the standard deviation of $\overline{X}_{\cdot j}^{1} - \overline{X}_{\cdot j}^{0}$ computed over 300 pure random allocations.

Figure 2 shows the distribution of standardized differences in each covariate (see Morgan and Rubin [16]) for São Paulo, the largest Brazilian municipality (18,182 sectors). It can be easily seen that differences are remarkably smaller for the haphazard allocations than for the rerandomization allocations, which, in turn, are remarkably smaller than for the pure randomization allocations. It is important to mention that this same pattern is verified in all other municipalities.

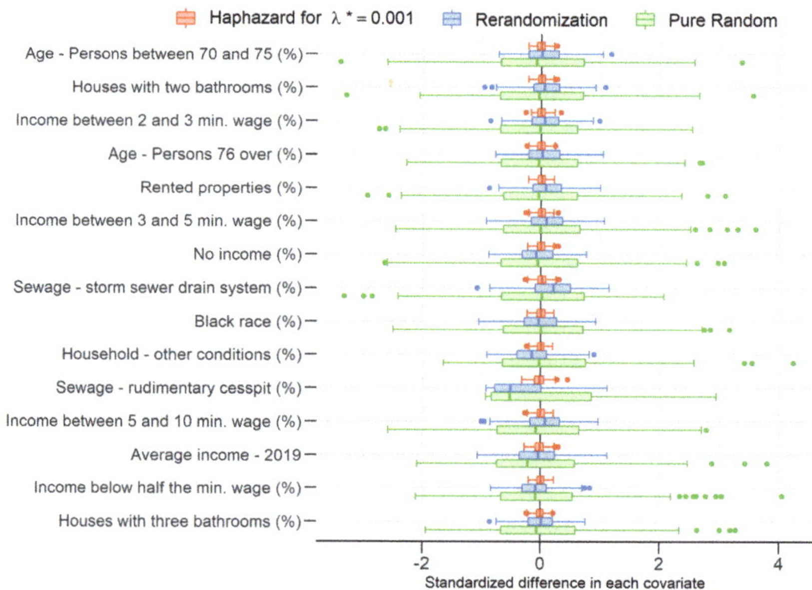

Figure 2. Difference between groups 1 (sampled sectors) and 0 (not sampled sectors) with respect to average standardized covariate values for each type of allocation.

4.2. Root Mean Square Errors of Simulated Estimations

We now consider simulated scenarios where, once we have sampled the sectors in each municipality, we estimate the municipality's SARS-CoV-2 prevalence based on observed prevalences on these sectors. Here, SARS-CoV-2 prevalence in each sector is simulated by the auxiliary regression model described in Section 3.1. To assess the estimation error yielded in each sampling method and to estimate variability, we compute, for each municipality, the root mean square error (RMSE) and the standard deviation of estimates, as follows:

$$RMSE(\hat{\theta}) = \sqrt{\frac{1}{r} \sum_{a=1}^{r} (\hat{\theta}_a - \theta)^2} \qquad SD(\hat{\theta}) = \sqrt{\frac{1}{r-1} \sum_{a=1}^{r} (\hat{\theta}_a - E(\hat{\theta}))^2}, \qquad (12)$$

where $r = 300$ denotes the number of allocations, $\hat{\theta}_a$ denotes the SARS-CoV-2 prevalence estimated from allocation a, θ denotes the SARS-CoV-2 prevalence considering all sectors of the municipality, and $E(\hat{\theta})$ denotes the average of $\hat{\theta}_a$ computed over r allocations.

Table 2 presents the $RMSE(\hat{\theta})$ and $SD(\hat{\theta})$ yielded by each sampling method for the 10 municipalities selected for this study. Both the Haphazard and the Rerandomization methods show RMSEs and SDs that are much smaller than the pure randomization method. Moreover, the Haphazard method outperforms the Rerandomization method, in the following sense: (a) the Haphazard method yields smaller RMSEs than the Rerandomization methods (in 9 out of 10 municipalities for this simulation); (b) moreover, the SDs are substantially smaller for the Haphazard method.

Table 2. Root mean square error (RMSE) and standard deviation (SD); red: best result; black: worst.

City	Haphazard		Rerandomizaton		Pure Randomizaton	
	RMSE	**SD**	**RMSE**	**SD**	**RMSE**	**SD**
São Paulo	1.6558%	1.6516%	2.4683%	2.3900%	4.9930%	4.9899%
Rorainópolis	0.8582%	0.7487%	1.5116%	1.4310%	3.0028%	3.0008%
Rio de Janeiro	1.3864%	1.3310%	1.9441%	1.9394%	4.6324%	4.6216%
Oiapoque	1.3887%	1.3835%	1.7651%	1.7509%	3.2107%	3.2107%
Marília	1.1624%	1.1603%	1.4787%	1.4737%	3.4950%	3.4919%
Iguatu	0.8329%	0.8196%	1.3029%	1.3025%	3.9094%	3.9003%
Cruzeiro do Sul	1.3873%	1.3489%	2.0482%	2.0457%	5.0029%	5.0003%
Corrente	0.7496%	0.7000%	1.0708%	1.0665%	2.8250%	2.8230%
Campos dos Goytacazes	0.9419%	0.9350%	1.8786%	1.8522%	4.4839%	4.4829%
Brasília	1.7978%	1.3434%	1.5739%	1.5299%	3.9608%	3.9539%

The RMSEs analyzed in the last paragraphs can be used to compute the sample size required to achieve a target precision in the statistical estimation of SARS-CoV-2 prevalence (as mentioned in Section 3, each sampling unit consists of a census sector containing around 200 households; the sample size refers to the number of sectors to be selected from each municipality). Figure 3 shows RMSEs as a function of sample size. If the sample size for each municipality is calibrated in order to achieve the target precision of the original Epicovid19 study (black horizontal line), using the Haphazard method implies an operating cost 40% lower than using the Rerandomization method and 80% lower than using pure randomization.

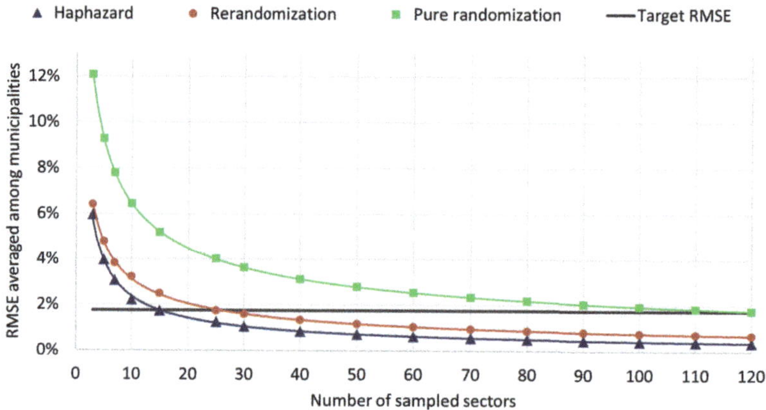

Figure 3. RMSE averaged among municipalities x number of sampled sectors.

5. Final Remarks

Both Haphazard and Rerandomization are promising methods in generating samples that provide good estimates of the population parameters with potentially reduced sample sizes and consequent operating costs. Alternatively, if we keep the same sample sizes, the use of Haphazard or Rerandomization methods will substantially improve the precision of statistical estimation. The Rerandomization method is simple to implement. The Haphazard intentional method requires the use of numerical optimization software and the empirical calibration of auxiliary parameters. Nevertheless, from a computational point of view, as the dimension of the covariate space or the number of elements to be allocated increases, the Haphazard method will be exponentially more efficient. Finally, the theoretical framework of the Rerandomization method has been fully developed [6,16]. In further research, we intend to better develop the theoretical framework of the Haphazard intentional sampling method and continue to show its potential for applied statistics.

Author Contributions: Conceptualization, M.M., J.S. and M.L.; Data acquisition, M.M.; Preprocessing and analysis, M.M., R.W. and M.L.; Programs implementation, M.L. and R.W.; Analysis of results, all authors. All authors have read and agreed to the published version of the manuscript.

Funding: This research was funded by CEPID-CeMEAI–Center for Mathematical Sciences Applied to Industry (grant 2013/07375-0, São Paulo Research Foundation–FAPESP), CEPID-RCGI–Research Centre for Gas Innovation (grant 2014/50279-4, São Paulo Research Foundation–FAPESP), and CNPq– the Brazilian National Council of Technological and Scientific Development (grant PQ 301206/2011-2).

Institutional Review Board Statement: Not applicable.

Informed Consent Statement: Not applicable.

Data Availability Statement: Raw data are available at: http://www.epicovid19brasil.org/?page_id=472 (accessed on 11 April 2021). Computer code is available at: https://github.com/marcelolauretto/Haphazard_MaxEnt2021 (accessed on 11 April 2021).

Conflicts of Interest: The authors declare no conflict of interest.

Abbreviations

The following abbreviations are used in this manuscript:

IBGE	Instituto Brasileiro de Geografia e Estatística (Brazilian Institute of Geography and Statistics)
IBOPE	Instituto Brasileiro de Opinião Pública e Estatística (Brazilian Institute of Public Opinion and Statistics)
MILP	Mixed-Integer Linear Programming
MIQP	Mixed-Integer Quadratic Programming
RMSE	Root mean square error
SD	Standard deviation

References

1. Lauretto, M.S.; Nakano, F.; Pereira, C.A.B.; Stern, J.M. Intentional Sampling by goal optimization with decoupling by stochastic perturbation. *AIP Conf. Proc.* **2012**, *1490*, 189–201.
2. Lauretto, M.S.; Stern, R.B.; Morgan, K.L.; Clark, M.H.; Stern, J.M. Haphazard intentional allocation an rerandomization to improve covariate balance in experiments. *AIP Conf. Proc* **2017**, *1853*, 050003.
3. Fossaluza, V.; Lauretto, M.S.; Pereira, C.A.B.; Stern, J.M. Combining Optimization and Randomization Approaches for the Design of Clinical Trials. In *Interdisciplinary Bayesian Statistics*; Springer: New York, NY, USA, 2015; pp. 173–184.
4. Stern, J.M. Decoupling, Sparsity, Randomization, and Objective Bayesian Inference. *Cybern. Hum. Knowing* **2008**, *15*, 49–68.
5. Lauretto, M.S.; Stern, R.B.; Ribeiro, C.O.; Stern, J.M. Haphazard Intentional Sampling Techniques in Network Design of Monitoring Stations. *Proceedings* **2019**, *33*, 12. [CrossRef]
6. Morgan, K.L.; Rubin, D.B. Rerandomization to improve covariate balance in experiments. *Ann. Stat.* **2012**, *40*, 1263–1282. [CrossRef]
7. Stern, J.M. Symmetry, Invariance and Ontology in Physics and Statistics. *Symmetry* **2011**, *3*, 611–635. [CrossRef]
8. Golub, G.H.; Van Loan, C.F. *Matrix Computations*; JHU Press: Baltimore, MD, USA, 2012.
9. Wolsey, L.A.; Nemhauser, G.L. *Integer and Combinatorial Optimization*; John Wiley & Sons: Hoboken, NJ, USA, 2014.
10. Ward, J.; Wendell, R. Technical Note-A New Norm for Measuring Distance Which Yields Linear Location Problems. *Oper. Res.* **1980**, *28*, 836–844. [CrossRef]
11. Murtagh, B.A. *Advanced Linear Programming: Computation And Practice*; McGraw-Hill International Book Co.: New York, NY, USA, 1981.
12. EPICOVID19. Available online: http://www.epicovid19brasil.org/?page_id=472 (accessed on 21 August 2020).
13. Fleiss, J.L. Measuring nominal scale agreement among many raters. *Am. Psychol. Assoc. (APA)* **1971**, *76*, 378–382 [CrossRef]
14. R Core Team. *R: A Language and Environment for Statistical Computing*; R Foundation for Statistical Computing: Vienna, Austria, 2020.
15. Gurobi Optimization Inc. *Gurobi: Gurobi Optimizer 9.01 Interface*; R Package Version 9.01; Gurobi Optimization Inc.: Beaverton, OR, USA, 2021.
16. Morgan, K.L.; Rubin, D.B. Rerandomization to Balance Tiers of Covariates. *J. Am. Stat. Assoc.* **2015**, *110*, 1412–1421. [CrossRef] [PubMed]

physical sciences forum

Proceeding Paper

Legendre Transformation and Information Geometry for the Maximum Entropy Theory of Ecology [†]

Pedro Pessoa

Department of Physics, University at Albany (SUNY), Albany, NY 12222, USA; ppessoa@albany.edu

† Presented at the 40th International Workshop on Bayesian Inference and Maximum Entropy Methods in Science and Engineering, online, 4–9 July 2021.

Abstract: Here I investigate some mathematical aspects of the maximum entropy theory of ecology (METE). In particular I address the geometrical structure of METE endowed by information geometry. As novel results, the macrostate entropy is calculated analytically by the Legendre transformation of the log-normalizer in METE. This result allows for the calculation of the metric terms in the information geometry arising from METE and, by consequence, the covariance matrix between METE variables.

Keywords: METE; metabolic rate distributions; information geometry; Legendre transformation; Lambert W function

Citation: Pessoa, P. Legendre Transformation and Information Geometry for the Maximum Entropy Theory of Ecology. *Phys. Sci. Forum* **2021**, *3*, 1. https://doi.org/10.3390/psf2021003001

Academic Editors: Wolfgang von der Linden and Sascha Ranftl

Published: 3 November 2021

Publisher's Note: MDPI stays neutral with regard to jurisdictional claims in published maps and institutional affiliations.

1. Introduction

The method of maximum entropy (MaxEnt) is usually associated with Jaynes' work [1–3] connecting statistical physics and the information entropy proposed by Shannon [4]—although its mathematics is known since Gibbs [5]. It consists of selecting probability distributions by maximizing a functional—namely entropy—usually under a set of expected values constraints, arriving at what is known as Gibbs distributions. Since Shore and Johnson [6] MaxEnt has been understood as a general method for inference—see also [7–9]—hence it is not surprising that (i) Gibbs distributions are what is known in statistical theory as exponential family—the only distributions for which sufficient statistics exist (see e.g., [10]), (ii) MaxEnt encompasses the methods of Bayesian statistics [11], and (iii) MaxEnt has found successful applications in several fields of science (e.g., [12–22]).

One of the scientific fields in which MaxEnt has been successfully applied is macroecology. The work of Harte and collaborators [23–27] presents what is known as the maximum entropy theory of ecology (METE). It consists of finding, through MaxEnt, a joint conditional distribution for the abundance of a species and the metabolic rate of its individuals. From the marginalization and expected values of the MaxEnt distribution, it is possible to obtain (i) the species abundance distribution (Fisher's log series), (ii) the species-area distribution, (iii) the distribution for metabolic rates over individuals, and (iv) the relationship between the metabolic rate of individuals in a species and that species abundance —for a comprehensive confirmation of METE with experimental data see [28]. In a recent article Harte [29] brings forward the need for dynamical models based on MaxEnt, as METE assume the variables to be static—It is relevant to say that Jaynes applied dynamical methods based on information theory for nonequilibrium statistical mechanics [30] leading to what is known as maximum caliber [31,32]. However, maximum caliber assumes a Hamiltonian dynamics and, therefore, does not generalize to ecology and other complex systems.

The field known as information geometry (IG) [33–36] assigns a Riemannian geometry structure to probability distributions. In information geometry the distances are given by the Fisher-Rao information metric (FRIM) [37,38], which is the only metric in accordance with the grouping property of probability distributions [39]. IG has found important applications for probabilistic dynamical systems [34,40–43]. Here the FRIM terms for the distributions arising from METE will be calculated. In a future publication I will evolve METE into entropic dynamical models for ecology, as explained in [43], in order to do so it is necessary to calculate the macrostate entropy and the FRIM terms—which can be obtained from differentiating the macrostate entropy. Therefore, present article performs the calculations necessary for an entropic dynamics model for macroecology.

The layout of the paper is as follows: The following section (2) presents MaxEnt in general terms followed by the MaxEnt process in METE. In particular we obtain the macrostate entropy through the Legendre transform, and the Lambert W special function [44,45], which is a novel result to the best of my knowledge. Section 3 presents some general results of IG and calculate the information metric terms for METE. Section 4 concludes the present article by commenting on possible applications and perspectives for IG in a dynamical theory of macroecology.

2. Maximum Entropy

In information theory, probability distributions encode the available information about a system's variables $x \in \mathcal{X}$. MaxEnt consists of updating from a prior distribution $q(x)$—usually, but not necessarily, taken to be uniform—to a posterior $\rho(x)$ that maximizes the entropy functional under a set of constraints meant to represent the known information about the system. Usually these constraints are the expected values A^i of a set of real valued functions $\{a^i(x)\}$ namely sufficient statistics. The distribution ρ is found as the solution to the following optimization problem

$$\max_{\rho} \quad H[\rho] = -\int dx\, \rho(x) \log\left(\frac{\rho(x)}{q(x)}\right), \tag{1a}$$

$$s.t. \quad \int dx\, \rho(x) = 1 \tag{1b}$$

$$\int dx\, a^i(x)\rho(x) = A^i. \tag{1c}$$

where $\int dx$ refers to the appropriate measure of the set $\mathcal{X}$; if one is interested in a discrete set $\mathcal{X} = \{x_\mu\}$, where μ corresponds to an enumeration of $\mathcal{X}$, we have $\int dx = \sum_\mu$, if one is interested in a continuous subset of real variables, e.g., $\mathcal{X} = [a,b]$, we have $\int dx = \int_a^b dx$.

The solution of (1) is the Gibbs distribution

$$\rho(x|\lambda_1, \lambda_2, ..., \lambda_n) = \frac{q(x)}{Z(\lambda)} \exp\left(-\sum_{i=1}^{n} \lambda_i a^i(x)\right), \tag{2}$$

where $\lambda = \{\lambda_i\}$ is the set of Lagrange multipliers dual to the expected values $A = \{A^i\}$ and $Z(\lambda)$ is a normalization factor given by

$$Z(\lambda) = \int dx\, q(x) \exp\left(-\lambda_i a^i(x)\right). \tag{3}$$

Above, and on the remainder of this article, we use Einstein's summation notation $A_i B^i = \sum_i A_i B^i$. The expected values can be recovered as

$$A^i = -\frac{1}{Z}\frac{\partial Z}{\partial \lambda_i} = \frac{\partial F}{\partial \lambda_i}, \quad \text{where} \quad F(\lambda) \doteq -\log(Z(\lambda)). \tag{4}$$

We will refer to F as the log-normalizer, which displays a role similar to free energy in statistical mechanics.

If one is able to invert the equations arriving from (4), obtaining this way $\lambda_i(A)$ they can express the probability distributions in terms of the expected values, $\rho(x|A) = \rho(x|\lambda(A))$. This also allows one to calculate the entropy H at its maximum—that means

$H[\rho(x|A)]$ for ρ in (2)—as a function of the expected values, rather than a functional of ρ, obtaining

$$H(A) \doteq H[\rho(x|\lambda(A))] = - \int dx\, \rho(x|\lambda(A)) \log \frac{\rho(x|\lambda(A))}{q(x)} = \lambda_i(A)A^i - F(\lambda(A)) \,. \quad (5)$$

We will refer to $H(A)$ as the macrostate entropy, which is what we refer to in statistical mechanics as thermodynamical entropy—meaning the one that appears in the laws of thermodynamics (Since the arguments that identify the macrostate entropy as the thermodynamical entropy assume that the sufficient statistics are conserved quantities in a Hamiltonian dynamics [2], analogous 'laws of thermodynamics' - e.g., conservation of A^2 in (12) or an impossibility of H in (15) to decrease— are not expected in ecological systems). One can see from (5) that $H(A)$ is the Legendre transformation [46] of $F(\lambda)$. It also follows that $\lambda_i = \frac{\partial H}{\partial A^i}$.

METE

The first step towards a MaxEnt description involves choosing the appropriate variables for the problem at hand. In METE [24] one assumes an ecosystem of S species supporting N individuals with a total metabolic rate E, meaning in a unit of time the ecosystem consumes a quantity E of energy. The state of the system x on MaxEnt is defined for a singular species as the number of individuals (abundance) n, $n \in \{1, 2, \ldots, N\}$ and the metabolic rate of an individual of that species ε, $\varepsilon \in [1, E]$—note that one can choose a system of units so that the smallest metabolic rate is the unit, $\varepsilon_{min} = 1$. We represent the state as $x = (n, \varepsilon)$.

The second step consists of assigning the sufficient statistics that appropriately captures the information about the system. In METE [24] the statistics chosen are the number of individuals in the species $a^1(n, \varepsilon) \doteq n$ and the total metabolic rate $a^2(n, \varepsilon) \doteq n\varepsilon$. Substituting these into the defined expected value constrains for the sufficient statistics (1), we obtain constraints on average abundance per species

$$A^1 = \sum_{n=1}^{N} \int_1^E d\varepsilon\, n\, \rho(n, \varepsilon|\lambda) = \frac{N}{S} \doteq N' \,, \quad (6)$$

and a constrain on the average metabolic consumption per species

$$A^2 = \sum_{n=1}^{N} \int_1^E d\varepsilon\, n\varepsilon\, \rho(n, \varepsilon|\lambda) = \frac{E}{S} \doteq E' \,. \quad (7)$$

The defined variable N' and E' will replace A^1 and A^2, respectively, when convenient.

Having the state variables and the sufficient statistics chosen, we can compute all quantities defined in the previous subsection for the specific system defined by METE. With a uniform prior q, justified by the fact that at its level of complexity organisms should be considered as distinguishable, this leads to the canonical distribution (2) of the form

$$\rho(n, \varepsilon|\lambda) = \frac{1}{Z(\lambda)} e^{-\lambda_1 n} e^{-\lambda_2 n\varepsilon} \,, \quad (8)$$

where the normalization factor (3) is given by

$$Z(\lambda) = \sum_{n=1}^{N} \int_1^E d\varepsilon\, e^{-\lambda_1 n} e^{-\lambda_2 n\varepsilon} = \sum_{n=1}^{N} e^{-\lambda_1 n} \left(\frac{e^{-\lambda_2 n} - e^{-\lambda_2 nE}}{\lambda_2\, n} \right) \,, \quad (9)$$

from which the expected values (4) can be calculated as

$$A^1 = N' = \frac{1}{\lambda_2\, Z(\lambda)} \sum_{n=1}^{N} e^{-\lambda_1 n}\left(e^{-\lambda_2 n} - e^{-\lambda_2 nE}\right), \tag{10a}$$

$$A^2 = E' = \frac{1}{\lambda_2}\left[1 + \frac{1}{Z(\lambda)} \sum_{n=1}^{N} e^{-\lambda_1 n}\left(e^{-\lambda_2 n} - Ee^{-\lambda_2 nE}\right)\right]. \tag{10b}$$

These are complicated equations, however some approximations may make them more treatable.

A fair assumption, knowing what the variables are supposed to represent, is that there are far more individuals than species, $N \gg S$ and the average metabolic rate per individual is far greater than the unit of metabolic rate $E/N = E'/N' \gg 1$. This allows for a sequence of approximation that we will treat like assumptions here, namely (i) $e^{-\lambda_2 nE} \ll e^{-\lambda_2 n}$, (ii) $Ee^{-\lambda_2 nE} \ll e^{-\lambda_2 n}$, (iii) $\lambda_1 + \lambda_2 \ll 1$, and (iv) $e^{-(\lambda_1+\lambda_2)N} \ll 1$. Further explanation on the validity of these assumptions, under $S \ll N \ll E$, can be seen in [24,26] and their confirmation by numerical calculation can be seen in [24]. Under this understanding we can substitute (9) into (10a) obtaining

$$N' = \frac{\sum_{n=1}^{N} e^{-\lambda_1 n}\left(e^{-\lambda_2 n} - e^{-\lambda_2 nE}\right)}{\sum_{n=1}^{N} \frac{1}{n} e^{-\lambda_1 n}\left(e^{-\lambda_2 n} - e^{-\lambda_2 nE}\right)} \approx \frac{\sum_{n=1}^{N} e^{-(\lambda_1+\lambda_2)n}}{\sum_{n=1}^{N} \frac{1}{n} e^{-(\lambda_1+\lambda_2)n}} \tag{11a}$$

$$N' \approx -\left[\frac{1}{(\lambda_1 + \lambda_2)\log(\lambda_1 + \lambda_2)}\right]. \tag{11b}$$

We can also rewrite (10b) obtaining

$$E' = \frac{1}{\lambda_2} + \frac{\sum_{n=1}^{N} e^{-\lambda_1 n}\left(e^{-\lambda_2 n} - Ee^{-\lambda_2 nE}\right)}{\sum_{n=1}^{N} \frac{1}{n} e^{-\lambda_1 n}\left(e^{-\lambda_2 n} - e^{-\lambda_2 nE}\right)} \approx \frac{1}{\lambda_2} + N'. \tag{12}$$

In order to obtain the macrostate entropy analytically (5) one needs to perform the Legendre transformation for METE, which includes inverting (11) and (12) obtaining $\lambda_1(N', E')$ and $\lambda_2(N', E')$. In page 149 of [24] it is said to be unfeasible. However, it is possible to do so obtaining

$$\lambda_1 = \beta(N') - \frac{1}{E' - N'}, \quad \text{and} \quad \lambda_2 = \frac{1}{E' - N'}, \tag{13}$$

where

$$\beta(N') \doteq -\left[N'\, W_{-1}\left(-\frac{1}{N'}\right)\right]^{-1}, \qquad \dot{\beta}(N') \doteq \frac{d\beta}{dN'} = \left[N'^2 - \frac{N'}{\beta(N')}\right]^{-1}, \tag{14}$$

and W_{-1} refers to the second main branch of the Lambert W function (see [44,45]). The details on how (13) inverts (11) and (12) are presented in Appendix A. The macrostate entropy can be calculated directly from (5) as

$$H(N', E') = N'\beta(N') + \log\left(E' - N'\right) - \log\left(N'\beta(N')\right) + 1. \tag{15}$$

With the calculation of the macrostate entropy finished, we can move into a geometric description of METE.

3. Information Geometry

This section presents the elementary notions of IG—for more in depth discussion and examples see e.g., [33–36]—and some useful identities for the IG of Gibbs distributions. IG consists of assigning a Riemmanian geometry structure to the space of probability distributions, meaning if a set of distributions $p(x|\theta)$ is parametrized by a finite number of coordinates, $\theta = \{\theta^i\}$, the distances—which are a measure of distinguishability—$d\ell$ between the neighbouring distributions $P(x|\theta + d\theta)$ and $P(x|\theta)$ are given by $d\ell^2 = g_{ij}d\theta^i d\theta^j$. The work of Cencov [39] demonstrated that the only metric invariant under Markov embeddings—and, therefore, the only one adequate to represent a space of probability distributions—is the metric of the form

$$g_{ij} = \int dx \, P(x|\theta) \frac{\partial \log P(x|\theta)}{\partial \theta^i} \frac{\partial \log P(x|\theta)}{\partial \theta^j} , \tag{16}$$

know as FRIM.

Considering the MaxEnt results presented in previous section, we can restrict our investigation to the Gibbs distributions using the expected values A as coordinates—$\theta^i = A^i$ and $P(x|\theta) = \rho(x|A)$ as in (2). Two useful expressions arise in that case—for proofs see e.g., [33]—first: the metric terms are the Hessian of the negative of macrostate entropy, meaning

$$g_{ij} = -\frac{\partial^2 H}{\partial A^i \partial A^j} = -\frac{\partial \lambda_i}{\partial A^j} , \tag{17}$$

and second: the covariance matrix between the sufficient statistics $a^i(x)$ is the inverse matrix of g_{ij}, meaning

$$C^{ij} g_{jk} = \delta^i_k , \quad \text{where} \quad C^{ij} = \left\langle a^i(x) a^j(x) \right\rangle - A^i A^j . \tag{18}$$

We can, then, see how these quantities are calculated for METE.

Information Geometry of METE

By substituting the macrostate entropy for METE (15) in (17) we obtain the FRIM terms:

$$g_{11} = -\dot{\beta}(N') + \frac{1}{(E' - N')^2} , \quad g_{12} = g_{21} = -\frac{1}{(E' - N')^2} ,$$
$$g_{22} = \frac{1}{(E' - N')^2} , \quad \text{and} \quad g = -\frac{\dot{\beta}(N')}{(E' - N')^2} . \tag{19}$$

where $g = \det g_{ij}$. Per (18) and from the general form of inverse matrix of a two dimensional matrix, the covariance matrix terms can be calculated directly inverting (19) obtaining

$$C^{11} = \frac{g_{22}}{g} = \frac{N'}{\beta(N')} - N'^2 , \quad C^{12} = C^{21} = -\frac{g_{12}}{g} = \frac{N'}{\beta(N')} - N'^2 ,$$
$$\text{and} \quad C^{22} = \frac{g_{11}}{g} = E'^2 - 2E'N' + \frac{N'}{\beta(N')} ; \tag{20}$$

completing the calculation. The matrix C^{ij} can be interpreted directly as the covariance between a species abundance and its total metabolic rate—METE sufficient statistics. The information metric terms presented in (19) allow for further studies on dynamical ecology from a information theory background, as we will comment in the following section.

4. Discussion and Perspectives

The present article calculates the macrostate entropy (15) for METE. This was made possible by the analytical calculation of the Lagrange multipliers (13) as functions of the expected values (10), previously believed to be unfeasible. This allows for a complete description of METE in terms of the average abundance N' and the expected metabolic rate E' of each of the ecosystem species. This opens a broad range of investigations possible by analytical calculations. In particular, the IG arising from METE is presented by calculating the FRIM terms in (19). Independently of any geometric interpretation, that was equivalent to calculate the covariance between METE sufficient statistics (20).

The variables that define an ecosystem's state are not expected to remain constant. Because of this, and the growing relevance of IG in dynamical systems, the calculations made in the present article are an important step into expanding maximum entropy ideas into further investigation in macroecology. The calculations done here allow for evolving METE into an entropic dynamics for ecology, as in the framework developed in [43], this venue of research will be explored in future publication.

Institutional Review Board Statement: This study did not involve humans nor other animals.

Informed Consent Statement: This study did not involve humans.

Data Availability Statement: No new data were created or analyzed in this study. Data sharing is not applicable to this article.

Acknowledgments: I would like to thank A. Caticha, J. Harte, E.A. Newman, and C. Camargo for insightful discussions.

Conflicts of Interest: The authors declare no conflict of interest.

Appendix A. On the Lambert W Function

In this appendix we will explain how (13) inverts (11) and (12). The Lambert W function is defined as the solution of

$$W(x)e^{W(x)} = x \,. \tag{A1}$$

The python library SciPy [47] implements the numerical calculation of W. This relates to (11b) in the following manner: by defining the variable $\beta = \lambda_1 + \lambda_2$ we obtain

$$\frac{1}{N'} = -\beta \log \beta \iff \frac{1}{\beta N'} e^{-\frac{1}{\beta N'}} = \frac{1}{N'} \,, \tag{A2}$$

hence $\beta = -\left[N' W\left(-\frac{1}{N'}\right)\right]^{-1}$. It is relevant to say that, from (A1), $W(x)$ is multivalued—the terminology Lambert W 'function' is used loosely. The several single-valued functions that solve (A1) are known as the different 'branches' of the Lambert W. In (13) and (14) only the W_{-1} branch was taken into account. Given our object of study, we will restrict to functions that are guaranteed to give a β that is real for large N'. As explained in [44], the two branches $W_0(x)$ and $W_{-1}(x)$ are real and analytic for $-e^{-1} < x < 0$, of equivalently β is real for $N' > e$. Coherent with the fact that (11) was derived for large N'.

Figure A1 presents the graphs of β obtained from the $W_0(x)$ and $W_{-1}(x)$ branches, as well as a comparison to the β obtained numerically from inverting (11a). Even though per (A2) the β obtained by both branches inverts (11b), it can be seen from Figure A1 that only the one obtained from $W_{-1}(x)$ approximates the inverse of (11a) for large N' and, therefore, it is the only one appropriate for the present investigation.

To complete the claim that λ_1 and λ_2 in (13) are calculated analytically, it is relevant to say that $W_{-1}\left(-\frac{1}{N'}\right)$ can be calculated using the series expansion (see page 153 in [44])

$$W_{-1}\left(-\frac{1}{N'}\right) = -\sum_{m=0}^{\infty} a_m z^m \,, \quad \text{where} \quad z = \sqrt{2(\log N' - 1)} \,, \tag{A3}$$

and a_m is defined recursively as $a_0 = 1$, $a_1 = 1$, and

$$a_m = \frac{1}{m+1}\left(a_{m-1} - \sum_{k=2}^{m-1} k\, a_k\, a_{m+1-k} \right). \tag{A4}$$

Note that real z implies $N' > e$, which is coherent with the condition for W_{-1} to be real.

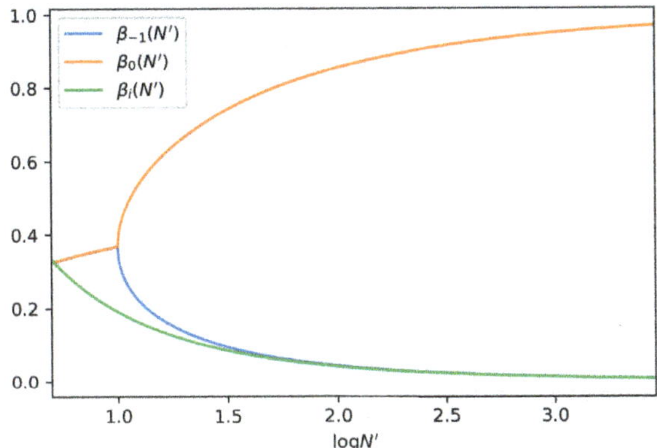

Figure A1. Graphical comparison between the functions defined as: $\beta_0(N') \doteq -\left[N'W_0\left(-\frac{1}{N'}\right)\right]^{-1}$, $\beta_{-1}(N') \doteq -\left[N'W_{-1}\left(-\frac{1}{N'}\right)\right]^{-1}$, and $\beta_i(N')$—obtained numerically from inverting (11a), here using $S = N/N' = 20$. W_0 and W_{-1} have complex values for $N' < e$, the graph above only plots the real part in that region.

References

1. Jaynes, E.T. Information theory and statistical mechanics. I. *Phys. Rev.* **1957**, *106*, 620. [CrossRef]
2. Jaynes, E.T. Gibbs vs Boltzmann entropies. *Am. J. Phys.* **1965**, *33*, 391–398. [CrossRef]
3. Jaynes, E.T. *Probability Theory: The Logic of Science*; Cambridge University Press: Cambridge, UK, 2003. [CrossRef]
4. Shannon, C.E. A Mathematical Theory of Communication. *Bell Syst. Tech. J.* **1948**, *27*, 379–423. [CrossRef]
5. Gibbs, J. *Elementary Principles in Statistical Mechanics*; Yale University Press: New Haven, CT, USA, 1902; reprinted by Ox Bow Press, Connecticut 1981.
6. Shore, J.; Johnson, R. Axiomatic derivation of the principle of maximum entropy and the principle of minimum cross-entropy. *IEEE Trans. Inf. Theory* **1980**, *26*, 26–37. [CrossRef]
7. Skilling, J. The Axioms of Maximum Entropy. In *Maximum-Entropy and Bayesian Methods in Science and Engineering*; Erickson, G.J., Smith, C.R., Eds.; Springer: Dordrecht, The Netherlands, 1988; Volumes 31–32. pp. 173–187. [CrossRef]
8. Caticha, A. Relative Entropy and Inductive Inference. In AIP Conference Proceedings; American Institute of Physics: College Park, MD, USA, 2004; Volume 707, pp. 75–96. [CrossRef]
9. Vanslette, K. Entropic Updating of Probabilities and Density Matrices. *Entropy* **2017**, *19*, 664. [CrossRef]
10. Daum, F. The Fisher-Darmois-Koopman-Pitman theorem for random processes. In Proceedings of the 25th IEEE Conference on Decision and Control, Athens, Greece, 10–12 December 1986; pp. 1043–1044. [CrossRef]
11. Caticha, A.; Giffin, A. Updating Probabilities. In *AIP Conference Proceedings*; American Institute of Physics: College Park, MD, USA, 2006; Volume 872, pp. 31–42. [CrossRef]
12. Golan, A. Information and Entropy Econometrics—A Review and Synthesis. *Found. Trends(R) Econom.* **2008**, *2*, 1–145. [CrossRef]
13. Bianconi, G. Entropy of network ensembles. *Phys. Rev. E* **2009**, *79*. [CrossRef]
14. Caticha, A.; Golan, A. An entropic framework for modeling economies. *Phys. A Stat. Mech. Its Appl.* **2014**, *408*, 149–163. [CrossRef]
15. Vicente, R.; Susemihl, A.; Jericó, J.P.; Caticha, N. Moral foundations in an interacting neural networks society: A statistical mechanics analysis. *Phys. A Stat. Mech. Its Appl.* **2014**, *400*, 124–138. [CrossRef]
16. Yong, N.; Ni, S.; Shen, S.; Ji, X. An understanding of human dynamics in urban subway traffic from the Maximum Entropy Principle. *Phys. A Stat. Mech. Its Appl.* **2016**, *456*, 222–227. [CrossRef]
17. Delgado-Bonal, A.; Martín-Torres, J. Human vision is determined based on information theory. *Sci. Rep.* **2016**, *6*. [CrossRef] [PubMed]
18. De Martino, A.; De Martino, D. An introduction to the maximum entropy approach and its application to inference problems in biology. *Heliyon* **2018**, *4*, e00596. [CrossRef]
19. Cimini, G.; Squartini, T.; Saracco, F.; Garlaschelli, D.; Gabrielli, A.; Caldarelli, G. The statistical physics of real-world networks. *Nat. Rev. Phys.* **2019**, *1*, 58–71. [CrossRef]

20. Dixit, P.D.; Lyashenko, E.; Niepel, M.; Vitkup, D. Maximum Entropy Framework for Predictive Inference of Cell Population Heterogeneity and Responses in Signaling Networks. *Cell Syst.* **2020**, *10*, 204–212.e8. [CrossRef]

21. Radicchi, F.; Krioukov, D.; Hartle, H.; Bianconi, G. Classical information theory of networks. *J. Physics Complex.* **2020**, *1*, 025001. [CrossRef]

22. Caldarelli, G.; Nicola, R.D.; Vigna, F.D.; Petrocchi, M.; Saracco, F. The role of bot squads in the political propaganda on Twitter. *Commun. Phys.* **2020**, *3*. [CrossRef]

23. Harte, J.; Zillio, T.; Conlisk, E.; Smith, A.B. Maximum entropy and the state-variable approach to macroecology. *Ecology* **2008**, *89*, 2700–2711. [CrossRef]

24. Harte, J. *Maximum Entropy and Ecology: A Theory of Abundance, Distribution, and Energetics*; Oxford University Press: Oxford, UK, 2011.

25. Harte, J.; Newman, E.A. Maximum information entropy: A foundation for ecological theory. *Trends Ecol. Evol.* **2014**, *29*, 384–389. [CrossRef]

26. Brummer, A.; Newman, E. Derivations of the Core Functions of the Maximum Entropy Theory of Ecology. *Entropy* **2019**, *21*, 712. [CrossRef]

27. Newman, E.A.; Wilber, M.Q.; Kopper, K.E.; Moritz, M.A.; Falk, D.A.; McKenzie, D.; Harte, J. Disturbance macroecology: A comparative study of community structure metrics in a high-severity disturbance regime. *Ecosphere* **2020**, *11*. [CrossRef]

28. Xiao, X.; McGlinn, D.J.; White, E.P. A Strong Test of the Maximum Entropy Theory of Ecology. *Am. Nat.* **2015**, *185*, E70–E80. [CrossRef] [PubMed]

29. Harte, J.; Umemura, K.; Brush, M. DynaMETE: A hybrid MaxEnt-plus-mechanism theory of dynamic macroecology. *Ecol. Lett.* **2021**. [CrossRef]

30. Jaynes, E.T. Where do we stand on maximum entropy? In *The Maximum Entropy Principle*; Levine, R.D., Tribus, M., Eds.; MIT Press: Cambridge, MA, USA, 1979. doi:10.1007/978-94-009-6581-2 10 [CrossRef]

31. Pressé, S.; Ghosh, K.; Lee, J.; Dill, K.A. Principles of maximum entropy and maximum caliber in statistical physics. *Reviews of Modern Physics* **2013**, 85, 1115–1141. revmodphys.85.1115 [CrossRef]

32. González, D.; Davis, S. The maximum caliber principle applied to continuous systems. *J. Phys. Conf. Ser.* **2016**, 720, 012006. 012006 . [CrossRef]

33. Caticha, A. The basics of information geometry. In *AIP Conference Proceedings*; American Institute of Physics: College Park, MD, USA, 2015, Volume 1641, pp. 15–26. [CrossRef]

34. Amari, S. *Information Geometry and Its Applications*; Springer: Berlin/Heidelberg, Germany, 2016. [CrossRef]

35. Ay, N.; Jost, J.; Lê, H.V.; Schwachhöfer, L. *Information Geometry*; Springer International Publishing: Berlin/Heidelberg, Germany, 2017. [CrossRef]

36. Nielsen, F. An Elementary Introduction to Information Geometry. *Entropy* **2020**, *22*, 1100. [CrossRef] [PubMed]

37. Fisher, R.A. Theory of Statistical Estimation. *Math. Proc. Camb. Philos. Soc.* **1925**, *22*, 700–725. [CrossRef]

38. Rao, C.R. Information and the accuracy attainable in the estimation of statistical parameters. *Bull. Calcutta Math. Soc.* **1945**, *37*, 81. [CrossRef]

39. Cencov, N.N. Statistical decision rules and optimal inference. In *Translations of Mathematical Monographs*; American Mathematical Society: Providence, RI, USA, 1981; Volume 53.

40. Hayashi, M.; Watanabe, S. Information geometry approach to parameter estimation in Markov chains. *Ann. Stat.* **2016**, *44*, 1495–1535. [CrossRef]

41. Felice, D.; Cafaro, C.; Mancini, S. Information geometric methods for complexity. *Chaos Interdiscip. J. Nonlinear Sci.* **2018**, *28*, 032101. [CrossRef]

42. Ruppeiner, G. Riemannian geometry in thermodynamic fluctuation theory. *Rev. Mod. Phys.* **1995**, *67*, 605. doi:10.1103/RevModPhys. 67.605. [CrossRef]

43. Pessoa, P.; Costa, F.X.; Caticha, A. Entropic dynamics on Gibbs statistical manifolds. *Entropy* **2021**, *23*, 494. [CrossRef] [PubMed]

44. Corless, R.M.; Jeffrey, D.J. The LambertW function. In *The Princeton Companion to Applied Mathematics*; Higham, N.J., Dennis, M.R., Glendinning, P., Martin, P.A., Santosa, F., Tanner, J., Eds.; Princeton University Press: Princeton, NJ, USA, 2016; Chapter III-17, pp. 151–155. [CrossRef]

45. Lehtonen, J. The Lambert W function in ecological and evolutionary models. *Methods Ecol. Evol.* **2016**, *7*, 1110–1118. [CrossRef]

46. Nielsen, F. Legendre Transformation and Information Geometry. Technical Report CIG-MEMO2, 2010. Available online: https://www2.sonycsl.co.jp/person/nielsen/Note-LegendreTransformation.pdf (accessed on 1 November 2021).

47. Lambert, W. SciPy Documentation. Available online: https://docs.scipy.org/doc/scipy/reference/generated/scipy.special. lambertw.html (accessed on 20 March 2021).

physical sciences forum

Proceeding Paper

Global Variance as a Utility Function in Bayesian Optimization [†]

Roland Preuss * and Udo von Toussaint

Max–Planck–Institut für Plasmaphysik, 85748 Garching, Germany; udt@ipp.mpg.de

* Correspondence: preuss@ipp.mpg.de

† Presented at the 40th International Workshop on Bayesian Inference and Maximum Entropy Methods in Science and Engineering, online, 4–9 July 2021.

Abstract: A Gaussian-process surrogate model based on already acquired data is employed to approximate an unknown target surface. In order to optimally locate the next function evaluations in parameter space a whole variety of utility functions are at one's disposal. However, good choice of a specific utility or a certain combination of them prepares the fastest way to determine a best surrogate surface or its extremum for lowest amount of additional data possible. In this paper, we propose to consider the global (integrated) variance as an utility function, i.e., to integrate the variance of the surrogate over a finite volume in parameter space. It turns out that this utility not only complements the tool set for fine tuning investigations in a region of interest but expedites the optimization procedure in toto.

Keywords: global optimization; Bayesian optimization; utility function; global variance

PACS: 02.50.-r; 52.65.-y

Citation: Preuss, R.; von Toussaint, U. Global Variance as a Utility Function in Bayesian Optimization. *Phys. Sci. Forum* **2021**, 3, 3. https://doi.org/10.3390/psf2021003003

Academic Editor: Sascha Ranftl

Published: 5 November 2021

Publisher's Note: MDPI stays neutral with regard to jurisdictional claims in published maps and institutional affiliations.

1. Introduction

In many experimental or theoretical approaches the effort of acquiring data may be costly, time consuming or both. The goal is to get insights in either the overall or extremal behaviour of a target quantity with respect to a set of parameters. If insights to functional dependencies between target and parameters are only to be obtained from a computationally expensive function, which may be considered as a black box function, it is instructive to employ surrogate modelling: already acquired data serve as a starting basis for establishing a surrogate surface in parameter space which then gets explored by Bayesian optimization [1]. An overall survey about Bayesian optimization may be found in [2], though it concentrates on an expected improvement (EI) utility and considers noise in the data only in the last paragraph, again by concentrating on EI. In contrast to this nice study we propose to alternate the different utilities at hand. Moreover, a fast information theory related to Bayesian optimization is shown in [3], though this approach approximates any black-box function by a parabolic form which differs from our ansatz letting the black-box function untouched. Interesting insights to multi-objective Bayesian optimization are provided by [4], which considers "multi-objective" in the sense of seeking the extrema–each is free of choice maximum or minimum–of a bunch of single-objective functions. However, the present paper concentrates on finding a common extremum depending on multiple parameters.

For the surrogate modelling we use the Gaussian process method (GP) [5] whose early stages date back to the middle of last century with very first efforts in geosciences [6] tackling the problem of surrogate modelling by so-called kriging [7]. Afterwards, GP has been appreciated much in the fields of neural networks and machine learning [8–12] and further work showed the applicability of active data selection via variance based criterions [13,14]. Our implementation of the GP method in this paper was already introduced at [15], and follows in notation–and apart from small amendments—the very instructive book of Rasmussen & Williams [5].

While in a previous work [16] we investigated the performance of utility functions for the expected improvement of an additional data point or for a data point with the maximal variance, in this paper we would like to introduce the global variance, i.e., the integral over the variance for a target surrogate within a region of interest with respect to a newly added data point. It is the substantial advantage of the Gaussian process method that such a task may be tackled simply on the basis of already acquired data, i.e., before new data have to be determined.

2. Global Variance for Gaussian Process-Based Model

In the following we concisely report the formulas leading to the results in this paper. For a thorough discussion of Gaussian processes please refer to the above mentioned papers, especially to the book of Rasmussen & Williams [5].

The problem of predicting function values in a multi-dimensional space supported by given data is a regression problem for a non-trivial function of unknown shape. Given are n target data y for input data vectors x_i of dimension $N_{\dim}$ with matrix $X = (x_1, x_2, ..., x_n)$ written as

$$
y = \begin{pmatrix} y_1 \\ y_2 \\ \vdots \\ y_n \end{pmatrix}, \quad
x_i = \begin{pmatrix} x_{i1} \\ x_{i2} \\ \vdots \\ x_{iN_{\dim}} \end{pmatrix}, \quad
X = \begin{pmatrix} x_{11} & x_{21} & \cdots & x_{n1} \\ x_{12} & x_{22} & \cdots & x_{n2} \\ \vdots & \vdots & \ddots & \vdots \\ x_{1N_{\dim}} & x_{2N_{\dim}} & \cdots & x_{nN_{\dim}} \end{pmatrix}.
\tag{1}
$$

We assume that target data y_i are blurred by Gaussian noise with $\sigma_{d_i}^2$. Further, we assume that the black box function interconnecting input X and target y is at least uniformly continuous and thereby justifies a description of a target surface with a surrogate from the Gaussian process method. Despite the experimental data and the physics background all quantities throughout this paper are without units.

The decisive quantity of a Gaussian process is the covariance function k describing the distance between two vectors x_p and x_q defined by

$$
k(x_p, x_q) = \sigma_f^2 \exp\left\{ -\frac{1}{2}\left| \frac{x_p - x_q}{\lambda} \right|^2 \right\}.
\tag{2}
$$

with the signal variance σ_f^2 and length scale λ. A covariance matrix $(K)_{ij} = k(x_i, x_j)$ considers the covariances of all input data X. The GP method describes a target value f_* at test input vector x_* by a normal distribution $p(f_*|X, y, x_*) \propto \mathcal{N}(\bar{f}_*, \text{var}(x_*))$ with mean $\bar{f}_* = k_*^T (K + \sigma_n^2 \Delta)^{-1} y$, and variance $\text{var}(x_*) = k(x_*, x_*) - k_*^T (K + \sigma_n^2 \Delta)^{-1} k_*$, where the term $\sigma_n^2 \Delta$ represents the degree of information in the data. While Δ is the diagonal matrix of the given variances $\sigma_{d_i}^2$, the variance σ_n^2 accounts for an overall noise in the data. Then the full covariance matrix M of the Gaussian Process is

$$
M = K + \sigma_n^2 \Delta
\tag{3}
$$

In Bayesian probability theory the three parameters $\theta = (\lambda, \sigma_f, \sigma_n)^T$ are considered to be hyper-parameters which show up in the marginal likelihood as

$$
\log p(y|\theta) = \text{const} - \frac{1}{2} y^T \left[K(\lambda, \sigma_f) + \sigma_n^2 \Delta \right]^{-1} y - \frac{1}{2} \log \left| K(\lambda, \sigma_f) + \sigma_n^2 \Delta \right|.
\tag{4}
$$

In [16], we showed, that for a sufficiently large data base the target surrogate is well described by using the expectation values of the hyper-parameters in the formulas for f_*

and $\mathrm{var}(f_*)$, at least well enough to determine a global optimum in a region of interest (RoI). The global optimum is found by employing utility functions, as there are the expected improvement $U_{\mathrm{EI}}(x_*) = \langle I \rangle = \int_{f_{\max}}^{\infty} f_* p(f_* | X, y, x_*) \mathrm{d}f_*$ and the maximum variance $U_{\mathrm{MV}}(x_*) = \mathrm{var}(f_*)$. For both the respective maximum at $x_{*\max} = \arg\max_{\{x_*\}} U_{\mathrm{EI/MV}}$ is sought. While the first one (U_{EI}) evaluates the possible information gain from a new data point, the second utility (U_{MV}) simply estimates the vector in input space with largest variance in the target surrogate.

In order to have a look on the implications of an additional data point in the surrogate, we propose a further utility function, i.e., the global variance defined on the multi-dimensional $\mathrm{RoI} \in [-1 : 1]$ by

$$U_{\mathrm{GV}} = \int_{-1}^{1} \mathrm{var}(x)\mathrm{d}x\,. \tag{5}$$

The exact integration shown in the Appendix A leads to

$$
\begin{aligned}
U_{\mathrm{GV}}^{\mathrm{exact}} &= 2^{N_{\dim}}\sigma_f^2 - \sigma_f^4\left(\frac{\sqrt{\pi}\lambda}{2}\right)^{N_{\dim}} \sum_{ij}^{n}\left(M^{-1}\right)_{ij} \\
&\quad \cdot \prod_{k}^{N_{\dim}}\left\{\mathrm{erf}\left[\frac{1}{\lambda} - \frac{x_{ik} + x_{jk}}{2\lambda}\right] - \mathrm{erf}\left[-\frac{1}{\lambda} - \frac{x_{ik} + x_{jk}}{2\lambda}\right]\right\}
\end{aligned}
\tag{6}
$$

Though Equation (6) represents the correct result, it may turn out in computation runs that the determination of the error-function is substantially time consuming compared to the total expenditure. Therefore, we would like to introduce two alternatives to the exact integration in Equation (5).

The first one is kind of an approximation. Since outside of the RoI the integrand in Equation (5) shows only trivial contributions we shift the upper and lower integral bounds to $\pm$ infinity and get from the simple Gaussian integrals

$$U_{\mathrm{GV}}^{\mathrm{inf}} \approx -\sigma_f^4\left(\sqrt{\pi}\lambda\right)^{N_{\dim}} \sum_{ij}^{n}\left(M^{-1}\right)_{ij} \exp\left\{-\frac{1}{4\lambda^2}(x_i - x_j)^T(x_i - x_j)\right\}\,. \tag{7}$$

We dropped the first term in the integral over $[\sigma_f^2 x]_{-\infty}^{\infty}$ for being infinity, since at least it is a constant contribution regardless of changes in the input X. Although the utility U_{GVinf} in Equation (7) is an approximation only, it has the advantage of being much easier accessible by numerical means and its computation performs much faster compared to Equation (6).

A second much more sophisticated approach is to insert an enveloping Gaussian function with adjustable location x_G (guiding center) and variance σ_G in the integral of Equation (5). Again the integration limits are shifted to $\pm$ infinity, however this time the enveloping Gaussian function takes care of the integrability and we get

$$
\begin{aligned}
U_{\mathrm{GV}}^{\mathrm{env}} &= \int_{-\infty}^{\infty} \mathrm{var}(x)\left(\frac{1}{\sqrt{2\pi\sigma_G^2}}\right)^{N_{\dim}} \exp\left[\frac{1}{2\sigma_G^2}(x - x_G)^T(x - x_G)\right]\mathrm{d}x \\
&= \sigma_f^2 - \sigma_f^4\left(\frac{\lambda}{\sqrt{2\sigma_G^2 + \lambda^2}}\right)^{N_{\dim}} \sum_{ij}^{n}\left(M^{-1}\right)_{ij} \\
&\quad \cdot \exp\left\{-\frac{1}{2}\left[\frac{x_i^T x_i + x_j^T x_j}{\lambda^2} + \frac{x_G^T x_G}{\sigma_G^2} - \frac{\left(\frac{x_i + x_j}{\lambda^2} + \frac{x_G}{\sigma_G^2}\right)^T\left(\frac{x_i + x_j}{\lambda^2} + \frac{x_G}{\sigma_G^2}\right)}{\frac{2}{\lambda^2} + \frac{1}{\sigma_G^2}}\right]\right\}\,.
\end{aligned}
\tag{8}
$$

The two parameters x_G and σ_G provide a toolset to guide the search for the next target data evaluations: a smaller standard deviation σ_G shifts the attention to the center of the enveloping Gaussian, while x_G gives us the possibility to focus on certain regions in the RoI.

All three utilities employing the global variance, U_{GV}^{exact}, U_{GV}^{inf}, U_{GV}^{env} require the inversion of the full covariance matrix M of Equation (3). Since the inversion has to be performed for every newly proposed test input x_* this is the main time consuming part in the whole procedure. Let us remind the reader, that the method we are proposing fully resides on input space (together with already acquired data) and the bottle neck is the generation of new target data. Therefore, the starting condition of very expensive (aka time consuming) data acquisition still holds. However, we can beneficially use blockwise matrix inversion [17] since a new input vector x_{n+1} expands the covariance matrix for one additional row and line only. Consequently, we reduced the computational effort to n^2-behaviour instead of n^3 for standard inversion.

3. Proof of Principle

We follow the global optimization scheme from Section 4 of [18]. Again we give proof of principle with a "black box" model featuring a broad parabolic maximum $2 - \sum_k^{N_{dim}} x_k^2 + (-1)^k 0.3$ together with a smaller cosine structure $0.1\cos[2\pi(x_k - 0.3)/\Delta_{cos}]$ on top of it, while we focus on a decent ripple on Δ_{cos}=0.6 in one and two dimensions (N_{dim}=1, 2).

Figures 1–3 show in left and right panels the results for one and two dimensions, respectively. The x axis to the right counts the number of newly acquired data for the utility comparisons in Figure 3 and in the bottom rows **(c)**, **(d)** of Figures 1 and 2.

For every newly added data point proposed by the various utilities, the distance between the true location of the global optimum and the maximal value of the surrogate residing on the data at hand is calculated in Figure 1. In a similar fashion, the search for the best surrogate description of the hidden model is shown in Figure 2.

Eventually Figure 3 demonstrates the use of an enveloping Gaussian function in the integral of the global variance by varying its center x_G, e.g., if an educated guess about the location of the extremal structure is at hand, i.e., the guiding center x_G is preset to the positive axis (1d) or the quadrant (2d) with the true model maximum. Consequently, Table 1 displays for 1d the specific number of data and for 2d the saturation level for which the target surrogate enters the stage of resembling the true model, i.e. the summed up (absolute) differences between all grid points of the target surrogate and the model starts to diminish with the number of target data only.

Table 1. Comparison of enveloping Gaussian utilities with different integral weights and guiding centers in finding the best surrogate. **1d:** Changing step to solution. **2d:** Saturation level of the solution.

Integral Weight of env. Gaussian within RoI	0.6	0.8	0.95
1d: Corresponding width of env. Gaussian σ_G	1.38	0.95	0.71
U_{GV} with $x_G = 0$	15	14	22
U_{GV} with $x_G = 0.5$	13	12	13
2d: Corresponding width of env. Gaussian σ_G	0.99	0.79	0.67
U_{GV} with x_G=(0;0)	0.23	0.61	0.21
U_{GV} with x_G=(0.5;-0.5)	0.06	0.14	0.05

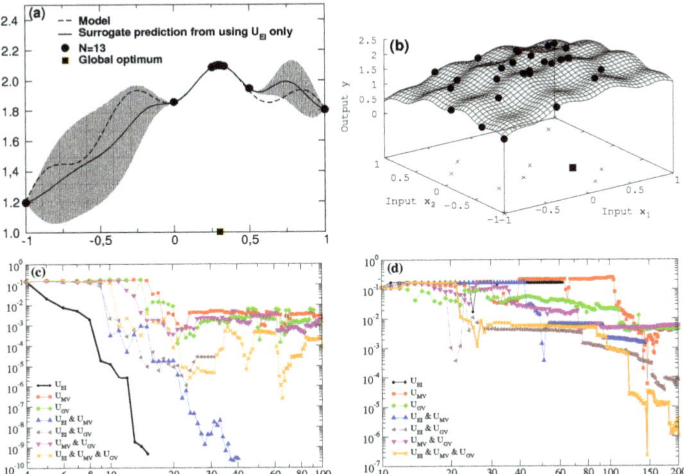

Figure 1. (**a**,**b**): One- and two-dimensional model with target data (full circles). The square in the bottom line/surface represents the true maximum. On the left the gray shaded area represents the uncertainty region of the surrogate (full line) from using the expected improvement utility only. On the right the points in the bottom surface are input data. Full circles represent additional data proposed by combination of all three utilities. (**c**,**d**): Distance surrogate/true maximum for different utilities employed in the global optimization procedure.

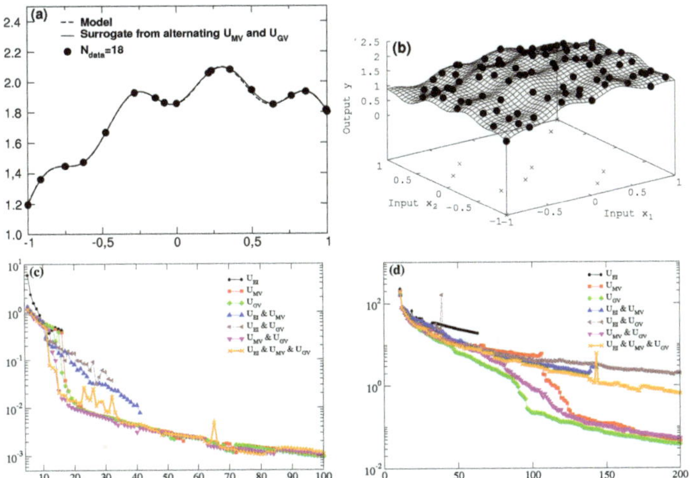

Figure 2. (**a**,**b**): One- and two-dimensional surrogate with newly acquired data. Surrogate solution (full line) on the left from using combination of U_{MV} and U_{GV}. Surrogate surface on the right from employing U_{GV} only. (**c**,**d**): Comparison of the differences between surrogate and true model integrated over RoI for various utilities.

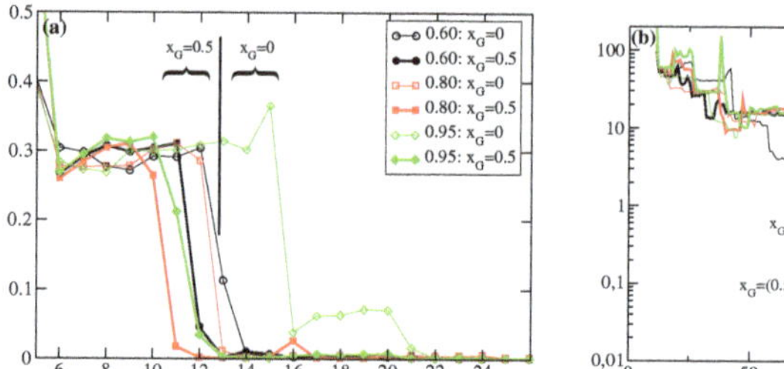

Figure 3. Summation over difference between grid points of target surrogate and true model as function of additionally acquired data. (**a**,**b**): One and two-dimensional case for enveloping Gaussian utility with various weights and guiding center at origin and at (0.5) or (0.5; −0.5).

4. Discussion

The results above show the usage of various utilities as a toolbox for surrogate modeling. Depending on the task—either to find an extremum or to get a best surrogate description of an unknown "black box" model—and depending on the prior knowledge at hand—presumption of location of the sought extremum or concentration on the region of interest—it is advisable to choose the most eligible utility function. However, even more promising is the combination of utilities of different character to profit from their benefits in toto and to compensate for pitfalls and drawbacks of one or the other utility.

As can be seen in the very first example in Figure 1 for the one-dimensional case the global optimum is found very fast with help of the expected improvement utility U_{EI} (starting to enter the bump with the correct extremum already below $N = 10$). However, a known drawback of this utility is that it gets stuck in local extrema and that it takes an unreasonably high number of additional data to get distracted from this pitfall.

This is taken into account for the two-dimensional case where the best result with lowest difference to the exact result is obtained by acting in combination of all three utilities U_{EI}, U_{MV} and U_{GV}. Focusing the maximum search on the utility regarding expected improvement alone (black line in Figure 1d would have got stuck in a local extremum with $y = 2.03$ in the "wrong" quadrant at $(−0.26; −0.29)$ for not recovering from this at all at about $N = 63$ (internal stop of algorithm for no improvement after entering computing accuracy level) and totally missing the true optimum with $y = 2.2$ at $(0.3; −0.3)$.

The situation changes for the task of getting a best overall description within the region of interest. To accomplish this the newly introduced global variance utility U_{GV} is of tremendous help both in one and two dimensions—either alone or in combination with at least the maximum variance utility U_{MV}. As shown in Figure 2d the best surrogate can already be established around ninety data points by employing U_{GV} only (full circles in the target surface of Figure 2b, with very few deviations from the true model left.

A guess about the approximate occurrence of an extremal structure–without excluding another region—can be emphasized by a further refinement to the global variance utility. In letting act an enveloping Gaussian within the global variance integral Equation (5) the result is not only much easier to be tackled from a computational point of view, but also the focus of the numerical search for the global optimum can be guided by predetermining the center of Gaussian x_G and its integral weight (aka width σ_{x_G}).

Figure 3 shows the results for three different integral weights (0.6; 0.8; 0.9) of the enveloping Gaussian function at two guiding centers: the first one at the origin corresponds to an ignorant scenario where one is not sure about a certain position of some global optimum at all. In the second approach we suppose that the extremal structure may be found in one dimension for positive values and thereby set $x_G = 0.5$, while in two dimensions it may be located within the quadrant with positive values for x_1 and negative ones for x_2 resulting in $x_G = [0.5; −0.5]$. As can be seen already in Figure 3, but all the more learned from the numbers of Table 1, displacing the center of the enveloping Gaussian function to the real center of the optimum of the hidden model facilitates the development of a best—regarding similarity to the true model—surrogate surface.

Author Contributions: The authors contributed equally to this work. All authors have read and agreed to the published version of the manuscript.

Funding: This research received no external funding.

Conflicts of Interest: The authors declare no conflict of interest.

Appendix A. Global Variance: Derivation of the Exact Integration

The variance at some (test) point $x^T = (x_1, x_2, \ldots, x_{N_{dim}})$ in a region confined to $[-1, 1]$ of dimension N_{dim} is

$$\operatorname{var}(x) = k(x, x) - k^T (K + \sigma_n^2 \Delta)^{-1} k \quad . \tag{A1}$$

The covariance $k(x_p, x_q)$ between pairs of input variables (x_p, x_q) is defined by

$$k(x_p, x_q) = \sigma_f^2 \exp\left[-\frac{(x_p - x_q)^T (x_p - x_q)}{2\lambda^2} \right] \quad . \tag{A2}$$

While the first term in Equation (A1) is simply $k(x, x) = \sigma_f^2$, we need for the second term

$$k = \sigma_f^2 \begin{pmatrix} \exp\left[-\frac{1}{2\lambda^2} (x - x_1)^T (x - x_1) \right] \\ \exp\left[-\frac{1}{2\lambda^2} (x - x_1)^T (x - x_2) \right] \\ \vdots \\ \exp\left[-\frac{1}{2\lambda^2} (x - x_1)^T (x - x_N) \right] \end{pmatrix} \tag{A3}$$

and the inversion of the matrix $M = K + \sigma_n^2 \Delta$, where the matrix elements are $\Delta_{ii} = \sigma_{d_i}$ ($\Delta_{ij} = 0$ for $i \neq j$) and $K_{ij} = \sigma_f^2 \exp\left[-\frac{1}{2\lambda^2} (x_i - x_j)^T (x_i - x_j) \right]$. For a given set of hyperparameters the matrix M does not depend on the test vector x and may be treated as a constant in integration over dx. Therefore, after the inversion has been performed, M^{-1} can easily be regarded as a pure number. So the second term in Equation (A1) is just a sum over all components with indices $\{i, j\} \in (1, \ldots, N)$,

$$k_*^T (K + \sigma_n^2 \Delta)^{-1} k_* = \sum_{i,j=1}^{N} k_{*i} \left(M^{-1} \right)_{ij} k_{*j} \tag{A4}$$

$$= \sigma_f^4 \sum_{i,j=1}^{N} \left(M^{-1} \right)_{ij} e^{ -\frac{(x - x_i)^T (x - x_i)}{2\lambda^2} } e^{ -\frac{(x - x_j)^T (x - x_j)}{2\lambda^2} } \quad . \tag{A5}$$

Further let us concentrate on the terms in the nominator of the exponential:

$$(x - x_i)^T (x - x_i) + (x - x_j)^T (x - x_j) = 2\left[-x^T \frac{x_i + x_j}{2} - \frac{x_i^T + x_j^T}{2} x \right] + x_i x_i^T + x_j x_j^T . \tag{A6}$$

Completing the square gives

$$2\left[\left(x - \frac{x_i + x_j}{2} \right)^T \left(x - \frac{x_i + x_j}{2} \right) \right] + \frac{1}{2} (x_i - x_j)^T (x_i - x_j) . \tag{A7}$$

We insert Equation (A7) in Equation (A5) and finally get for the variance

$$\text{var}(x) = \sigma_f^2 - \sigma_f^4 \sum_{i,j=1}^{N} \left(M^{-1} \right)_{ij} e^{-\frac{1}{4\lambda^2}(x_i-x_j)^T(x_i-x_j)} e^{-\frac{1}{\lambda^2}\left[\left(x-\frac{x_i+x_j}{2}\right)^T\left(x-\frac{x_i+x_j}{2}\right)\right]} . \tag{A8}$$

Only the second exponential in Equation (A8) depends on x and therefore needs to be considered in the integral of the global variance:

$$\int_{-1}^{1} dx \, \text{var}(x) . \tag{A9}$$

We insert Equation (A8) into Equation (A9) and let the integral stay only for the term with x dependency:

$$\int_{-1}^{1} dx \, \text{var}(x) = 2^{N_D}\sigma_f^2 - \sigma_f^4 \sum_{i,j=1}^{N} \left(M^{-1} \right)_{ij} e^{-\frac{1}{4\lambda^2}(x_i-x_j)^T(x_i-x_j)}$$
$$\cdot \int_{-1}^{1} dx \, e^{-\frac{1}{\lambda^2}\left[\left(x-\frac{x_i+x_j}{2}\right)^T\left(x-\frac{x_i+x_j}{2}\right)\right]} . \tag{A10}$$

Since the term in the exponential is quadratic it separates into a sum, which itself facilitates the separation of the integral into each dimension. Being simplified to a number of N_d one-dimensional integrals they can easily be solved by employing the error function. To prove this, let us have a closer look at the integral only:

$$\int_{-1}^{1} dx \, e^{-\frac{1}{\lambda^2}\left[\left(x-\frac{x_i+x_j}{2}\right)^T\left(x-\frac{x_i+x_j}{2}\right)\right]} = \prod_{k}^{N_{dim}} \int_{-1}^{1} dx_k e^{-\frac{1}{\lambda^2}\left[\left(x_k-\frac{x_{ik}+x_{jk}}{2}\right)^2\right]} . \tag{A11}$$

Focusing on a the kth integral and substituting $\tau_k = (x_k - \frac{x_{ik}+x_{jk}}{2})/\lambda$ some error functions evolve to end up finally in:

$$\int_{-1}^{1} dx_k e^{-\frac{1}{\lambda^2}\left[\left(x_k-\frac{x_{ik}+x_{jk}}{2}\right)^2\right]} = \lambda \int_{-\frac{1}{\lambda}-\frac{x_{ik}+x_{jk}}{2\lambda}}^{\frac{1}{\lambda}-\frac{x_{ik}+x_{jk}}{2\lambda}} d\tau_k e^{-\tau_k^2} \tag{A12}$$

$$= \lambda \left[\int_{\frac{1}{\lambda}-\frac{x_{ik}+x_{jk}}{2\lambda}}^{0} d\tau_k e^{-\tau_k^2} + \int_{0}^{-\frac{1}{\lambda}-\frac{x_{ik}+x_{jk}}{2\lambda}} d\tau_k e^{-\tau_k^2} \right] \tag{A13}$$

$$= \frac{\sqrt{\pi}}{2}\lambda \left\{ \text{erf}\left[\frac{1}{\lambda} - \frac{x_{ik}+x_{jk}}{2\lambda}\right] - \text{erf}\left[-\frac{1}{\lambda} - \frac{x_{ik}+x_{jk}}{2\lambda}\right] \right\} . \tag{A14}$$

This concludes the study. Simply inserting Equation (A14) into Equation (A10) succeeds in the result reported in the paper:

$$\int_{-1}^{1} dx \, \text{var}(x) = 2^{N_D}\sigma_f^2 - \sigma_f^4 \left(\frac{\sqrt{\pi}}{2}\lambda\right)^{N_{dim}} \sum_{i,j=1}^{N} \left(M^{-1} \right)_{ij} e^{-\frac{1}{4\lambda^2}(x_i-x_j)^T(x_i-x_j)}$$
$$\cdot \prod_{k}^{N_{dim}} \left\{ \text{erf}\left[\frac{1}{\lambda} - \frac{x_{ik}+x_{jk}}{2\lambda}\right] - \text{erf}\left[-\frac{1}{\lambda} - \frac{x_{ik}+x_{jk}}{2\lambda}\right] \right\} . \tag{A15}$$

References

1. Mockus, J. *Bayesian Approach to Global Optimization*; Springer: Berlin/Heidelberg, Germany, 1989.
2. Frazier, P.I. A Tutorial on Bayesian Optimization. 2018. Available online: http://xxx.lanl.gov/abs/1807.02811 (accessed on 11 September 2021).

3. Ru, B.; McLeod, M.; Granziel, D.; Osborne, M.A. Fast Information-theoretic Bayesian Optimisation. In Proceedings of the 35th International Conference on Machine Learning, Stockholm, Sweden, 10–15 July 2018; Volume 80.
4. Hernández-Lobato, D.; Hernández-Lobato, J.; Shah, A.; Adams, R.P. Predictive Entropy Search for Multi-objective Bayesian Optimization. In Proceedings of the 33rd International Conference on Machine Learning, New York, NY, USA, 20–22 June 2016; Volume 48.
5. Rasmussen, C.; Williams, C. *Gaussian Processes for Machine Learning*; MIT Press: Cambridge, UK, 2006.
6. Krige, D.G. A Statistical Approach to Some Basic Mine Valuation Problems on the Witwatersrand. *J. Chem. Metal. Mining Soc. S. Afr.* **1951**, *52*, 119–139.
7. Matheron, G. Principles of geostatistics. *Econ. Geol.* **1963**, *58*, 1246–1266. [CrossRef]
8. Barber, D. *Bayesian Reasoning and Machine Learning*; Cambridge University Press: Cambridge, UK, 2012.
9. Bishop, C. *Neural Networks for Pattern Recognition*; Oxford University Press: Oxford, UK, 1996.
10. Cohn, D. Neural Network Exploration Using Optimal Experiment Design. *Neural Netw.* **1996**, *9*, 1071–1083. [CrossRef]
11. MacKay, D.J.C. *Bayesian Approach to Global Optimization: Theory and Applications*; Kluwer Academic: Dordrecht, The Netherlands, 2013.
12. Neal, R.M. *Monte Carlo Implementation of Gaussian Process Models for Bayesian Regression and Classification*; Technical Report 9702; Department of Statistics, University of Toronto: Toronto, ON, Canada, 1997.
13. Seo, S.; Wallat, M.; Graepel, T.; Obermayer, K. Gaussian process regression: active data selection and test point rejection. In Proceedings of the International Joint Conference on Neural Networks, Como, Italy, 24–27 July 2000; pp. 241–246.
14. Gramacy, R.B.; Lee, H.K.H. Adaptive Design and Analysis of Supercomputer Experiments. *Technometrics* **2009**, *51*, 130–145. [CrossRef]
15. Preuss, R.; von Toussaint, U. Prediction of Plasma Simulation Data with the Gaussian Process Method. *Bayesian Inference and Maximum Entropy Methods in Science and Engineering*; Niven, R., Ed.; AIP Publishing: Melville, NY, USA, 2014; Volume 1636, p. 118.
16. Preuss, R.; von Toussaint, U. Global Optimization Employing Gaussian Process-Based Bayesian Surrogates. *Entropy* **2018**, *20*, 201. [CrossRef] [PubMed]
17. Invertible Matrix, Section 3.7: Blockwise Inversion. Available online: https://en.wikipedia.org/wiki/Invertible_matrix#Blockwise_inversion (accessed on 25 May 2021).
18. Preuss, R.; von Toussaint, U. Optimization Employing Gaussian Process-Based Surrogates. *Bayesian Inference and Maximum Entropy Methods in Science and Engineering*; Springer International Publishing: Berlin/Heidelberg, Germany, 2018; Volume 239.

 *physical sciences forum*

 MDPI

Proceeding Paper

Bayesian Surrogate Analysis and Uncertainty Propagation †

Sascha Ranftl * and Wolfgang von der Linden

Institute of Theoretical Physics-Computational Physics, Graz University of Technology, Petersgasse 16,
8010 Graz, Austria; vonderlinden@tugraz.at
* Correspondence: ranftl@tugraz.at
† Presented at the 40th International Workshop on Bayesian Inference and Maximum Entropy Methods in
Science and Engineering, online, 4–9 July 2021.

Abstract: The quantification of uncertainties of computer simulations due to input parameter uncertainties is paramount to assess a model's credibility. For computationally expensive simulations, this is often feasible only via surrogate models that are learned from a small set of simulation samples. The surrogate models are commonly chosen and deemed trustworthy based on heuristic measures, and substituted for the simulation in order to approximately propagate the simulation input uncertainties to the simulation output. In the process, the contribution of the uncertainties of the surrogate itself to the simulation output uncertainties is usually neglected. In this work, we specifically address the case of doubtful surrogate trustworthiness, i.e., non-negligible surrogate uncertainties. We find that Bayesian probability theory yields a natural measure of surrogate trustworthiness, and that surrogate uncertainties can easily be included in simulation output uncertainties. For a Gaussian likelihood for the simulation data, with unknown surrogate variance and given a generalized linear surrogate model, the resulting formulas reduce to simple matrix multiplications. The framework contains Polynomial Chaos Expansions as a special case, and is easily extended to Gaussian Process Regression. Additionally, we show a simple way to implicitly include spatio-temporal correlations. Lastly, we demonstrate a numerical example where surrogate uncertainties are in part negligible and in part non-negligible.

Keywords: uncertainty quantification; uncertainty propagation; surrogate models; meta-modeling; Bayesian

Citation: Ranftl, S.; von der Linden, W. Bayesian Surrogate Analysis and Uncertainty Propagation. *Phys. Sci. Forum* **2021**, *3*, 6. https://doi.org/10.3390/psf2021003006

Academic Editor: MaxEnt 2021 Scientific Committee

Published: 13 November 2021

Publisher's Note: MDPI stays neutral with regard to jurisdictional claims in published maps and institutional affiliations.

1. Introduction

Uncertainty quantification of simulations has gained increasing attention, e.g., in the field of Computational Engineering, in order to address doubtful parameter choices and assess the models' credibility. Surrogate models have become a popular tool to propagate simulation input uncertainties to the simulation output, particularly for modern day applications with high computational cost and many uncertain model parameters. For that, a parametrized surrogate model (synonyms: meta-model, emulator) is learned from a finite set of simulation samples. i.e., the surrogate is a function of the uncertain simulation input parameters that is 'fitted' to the simulation output data. The quality of this fit is then judged by heuristic diagnostics, and the surrogate deemed trustworthy, respectively. A key aspect of this procedure is that the surrogate can be evaluated much faster than the simulation, and still retains a reasonable approximation to the simulation.

The simulation is then substituted with the surrogate model in order to compute the marginal probability density function of the simulation output. The simulation uncertainties are thus inferred from the surrogate model instead of the original simulation model at a significantly reduced computational effort. While this practice allows for obtaining estimates on uncertainties of expensive simulations in the first place, the contribution of the uncertainty of the surrogate itself to the total simulation uncertainty is commonly neglected. In other words, the estimation of the surrogate parameters based on the finite set of simulation samples entails an additional uncertainty in the sought-for uncertainty of the simulation output. The purpose of this paper is to investigate this surrogate uncertainty as the natural measure for the surrogate's trustworthiness, and how the surrogate uncertainty affects the simulation output uncertainty.

In many cases, the surrogate uncertainty is indeed small if the heuristic diagnostics naively imply so. If the heuristic diagnostics imply that the surrogate is not trustworthy, one may resort to two options: (i) Acquire more simulation data until the surrogate is trustworthy. This is limited by the computational budget and the surrogate's convergence properties. (ii) Shrink the parameter space, e.g., omit a number of uncertain simulation parameters by assuming definite parameter values. However, in some cases, (i) is not feasible and (ii) is not desired. In this contribution, we demonstrate how to include surrogate uncertainties if the user deals with a surrogate model with doubtful trustworthiness.

Popular surrogate models are Polynomial Chaos Expansions [1–3] and Gaussian Process Regression [4,5], the latter of which has had its renaissance recently from within the machine learning community. In this work, we assume a Gaussian likelihood for the simulation data with unknown variance and given a generalized linear surrogate model (i.e., linear in the surrogate parameters) that includes Polynomial Chaos Expansions as a special case and is easily extended to Gaussian Process Regression. Other Bayesian perspectives on Uncertainty Quantification of computer simulations with these popular surrogate models are given in [6–13]. A comprehensive collection of reviews on Uncertainty Quantification, from the point of view of computational engineering and applied mathematics, can be found in [14]. In [7,15], a statistician's perspective is discussed. Here, we will use Bayesian Probability Theory [16].

2. Bayesian Uncertainty Quantification

We start with the general structure of uncertainty propagation problems based on surrogate models in Section 2.1. In Section 2.2, we analyze a generalized linear surrogate model with a Gaussian likelihood for the simulation data with unknown surrogate variance. In Section 2.3, we proceed to use the surrogate model to propagate the input uncertainties to the output, and show how the surrogate's uncertainties too can be included.

2.1. General Structure of the Problem

The goal in this paper is to quantify the uncertainties of the simulation results for the observable $z^{(x)}$ at different measurement points $x = 1, ..., N_x$ in the simulation domain. e.g., z could be the mechanical stress resulting from a structural analysis with a finite element simulation, where x could denote the location of the measurement probe in or on the analyzed structure. $z^{(x)}$ depends on unknown or uncertain model parameters $\boldsymbol{a} = \{a_i\}_{i=1}^{N_a}$, which are generally inferred from experimental data $\boldsymbol{d}_{\mathrm{exp}}$. Based on these data, Bayes' theorem allows for determining the posterior probability density function (pdf) for $\boldsymbol{a}$,

$$p(\boldsymbol{a} \mid \boldsymbol{d}_{\mathrm{exp}}, \mathcal{I}) \,, \tag{1}$$

where all background information on the experiment is subsumed in $\mathcal{I}$. The implications of the background information $\mathcal{I}$ will be discussed later. This pdf will be assumed to be (almost) arbitrary but given in the following considerations. It usually is the result of a statistical data analysis of the foregoing experiment. This experiment could be the measurement of some material property needed for the simulation, e.g., viscosity for a computational fluid dynamics simulation. The uncertainty of the model parameters $\boldsymbol{a}$ entails an uncertainty in the simulated observable $z^{(x)}$, and the latter is determined by the marginalization rule,

$$p(z^{(x)} \mid \boldsymbol{d}_{\mathrm{exp}}, \mathcal{I}) = \int p(z^{(x)} \mid \boldsymbol{a}, \boldsymbol{d}_{\mathrm{exp}}, \mathcal{I}) p(\boldsymbol{a} \mid \boldsymbol{d}_{\mathrm{exp}}, \mathcal{I}) dV_{\boldsymbol{a}} \,. \tag{2}$$

In the first pdf, we have struck out $d_{\exp}$ because the knowledge of a suffices to perform the simulation to obtain $z^{(x)}$. If a consists of only one or two parameters ($N_a = 1, 2$), then the numerical evaluation of the integral over the model parameters a will typically require a few dozens to hundreds of simulations. The uncertainty propagation is then done and no surrogate is needed. However, this is the trivial case, and usually a will consist of way more parameters. Let us assume that a consists of, e.g., four parameters. That would imply the need for performing simulations at least 10^5 times, which is way too CPU expensive for most real problems. This can be avoided if the simulations are replaced by a surrogate model that approximates the observable z by a suitable parametrized surrogate function $z_{\text{sur}} = g(a|c)$, where c are yet unknown parameters. The simulation may yield the observable $z^{(x)}$ at different sites x in the domain; however, x could also denote the time-instance in non-static problems. Clearly, the parameters of the surrogate model will also depend on those positions. Thus, we actually have

$$z^{(x)} \approx z_{\text{sur}}^{(x)} = g\left(a|c^{(x)}\right) . \tag{3}$$

The unknown parameters will be inferred from a suitable training data. To this end, simulations are performed for a finite set of model parameters $A_s = \{a_s^{(i)}\}_{i=1}^{N_s}$ and the corresponding observables $Z_s = \{z_s^{(x),(i)}\}_{i,x=1}^{N_s,N_x}$ are computed and combined in $D_{\text{sim}} = \{A_s, Z_s\}$. The surrogate parameters $c^{(x)}$ are then inferred from D_{sim}, and the surrogate is thus constructed. We now proceed to substitute the simulation for the surrogate, $z^{(x)} \to z_{\text{sur}}^{(x)}$, in order to solve Equation (2) at a significantly reduced computational cost. This implies that the background information has changed. We will denote this as $\tilde{\mathcal{I}}$, suggesting that we take the observable z entering the integral in Equation (2) from the surrogate model Equation (3) rather than from the expensive simulation. More precisely, instead of Equation (2), we now need to consider

$$p\left(z^{(x)} \mid d_{\exp}, D_{\text{sim}}, \tilde{\mathcal{I}}\right) = \int dV_a \, p\left(z^{(x)} \mid a, D_{\text{sim}}, \cancel{d_{\exp}}, \tilde{\mathcal{I}}\right) p\left(a \mid \cancel{D_{\text{sim}}}, d_{\exp}, \tilde{\mathcal{I}}\right) . \tag{4}$$

As far as the (second) pdf for the model parameters is concerned, we can omit the information on the training set D_{sim}, as it does not tell us anything about the model parameters. This pdf is actually the same as that in Equation (1), i.e., $p\left(a \mid d_{\exp}, \tilde{\mathcal{I}}\right) = p\left(a \mid d_{\exp}, \mathcal{I}\right)$, as it makes no difference for the model parameters how we solve the equations underlying the simulation. In the first pdf, we can omit the information on the experiment $d_{\exp}$, as we only need the simulation data D_{sim} to fix the surrogate model, which in turn defines the observable z. The first pdf can be further specified by the marginalization rule upon introducing the surrogate parameters $C = \{c^{(x)}\}_{x=1}^{N_x}$, where x is an index denoting the measurement points as introduced for $z^{(x)}$,

$$p\left(z^{(x)} \mid a, D_{\text{sim}}, \tilde{\mathcal{I}}\right) = \int dV_C \, p\left(z^{(x)} \mid C, a, \cancel{D_{\text{sim}}}, \tilde{\mathcal{I}}\right) p\left(C \mid \cancel{a}, D_{\text{sim}}, \tilde{\mathcal{I}}\right) .$$

The first pdf is uniquely fixed by the knowledge of a and C, hence D_{sim} is superfluous. Similarly, in the second pdf, where C is inferred from the training data, additional model parameters without the corresponding observables' values z, are useless. In summary, substituting the latter equation into Equation (4), we have

$$p\left(z^{(x)} \mid d_{\exp}, D_{\text{sim}}, \tilde{\mathcal{I}}\right) = \iint dV_a dV_C \, p\left(z^{(x)} \mid C, a, \tilde{\mathcal{I}}\right) p\left(C \mid D_{\text{sim}}, \tilde{\mathcal{I}}\right) p\left(a \mid d_{\exp}, \mathcal{I}\right) . \tag{5}$$

The first pdf is rather simple. According to the background information $\tilde{\mathcal{I}}$, we will determine the observable via the surrogate model. Since the necessary parameters $c^{(x)} \in C$ and a are part of the conditional complex, the surrogate model allows only one value

$$z_{\text{sur}}^{(x)} = g(a \mid c^{(x)}) \,.$$

for the observable. This means that $p(z^{(x)} \mid C, a, \tilde{\mathcal{I}})$ is equivalent to the probability density function for $z^{(x)}$ given $z^{(x)} = g(a \mid c^{(x)})$. Hence, the pdf is a Dirac-delta distribution

$$p(z^{(x)} \mid C, a, \tilde{\mathcal{I}}) = \delta\left[z^{(x)} - g(a \mid c^{(x)})\right] \,. \tag{6}$$

Finally, we have

$$p(z^{(x)} \mid d_{\text{exp}}, D_{\text{sim}}, \tilde{\mathcal{I}}) = \iint dV_a dV_C \, \delta\left[z^{(x)} - g(a \mid c^{(x)})\right] \, p(C \mid D_{\text{sim}}, \tilde{\mathcal{I}}) \, p(a \mid d_{\text{exp}}, \mathcal{I}) \,. \tag{7}$$

Before we can evaluate this integral, we first need to determine the two terms, which have their own independent significance. The last term is the result of a data analysis of a specific foregoing experiment, and will therefore not be treated here. We will suppress the background information in the following, as ambiguities should no longer occur.

2.2. Bayesian Analysis and Selection of the Surrogate Model

We recall that Equation (7) allows for determining the pdf for the observable based on the pdf for the model parameters, and the pdf for the parameters of the surrogate model $p(C \mid D_{\text{sim}})$ that we will determine now. To this end, we have to specify the form of the surrogate model. We use the expansion

$$z_{\text{sur}} = \sum_{\nu=1}^{N_p} c_\nu \Phi_\nu(a) \,, \tag{8}$$

in terms of basis functions $\Phi_\nu(a)$ and expansion coefficients c_ν. No further specification is needed at this point. Without loss of generality, we will use multi-variate Legendre polynomials for the numerical examples. This expansion is similar to the frequently used generalized Polynomial Chaos Expansfion [1], where the polynomials $\Phi_\nu(a)$ are orthogonal with respect to the L^2 inner product with the prior of a, $p(a)$, as an integration measure. However, here we actually consider a posterior $p(a \mid d_{exp})$ that generally has no standard form, for which no standard orthogonal polynomial chaos basis is known, and for which conditional independence of the model parameters a does not hold. Polynomial Chaos Expansions have been extended to arbitrary probability measures [17] and dependent parameters [18], but, in the present context; however, these polynomials are not of primary interest and would only complicate the numerical evaluation. Note that the approach presented here does not demand any orthogonality properties for the basis, and thus avoid the practical problems encountered with the construction of such orthogonal bases. Parameters here may have complex dependence structures, and the only requirement for the probability distribution $p(a \mid d_{\text{exp}})$ is that the integrals with respect to it exist. As outlined in Section 2.1, N_s simulations are performed for a set of model parameters $A_s = \{a_s^{(i)}\}_{i=1}^{N_s}$ and the corresponding observables Z_s are computed. The theory is so far agnostic to the experimental design of these simulations, and it is therefore not of concern here. Now, we want to determine the pdf for the surrogate parameters C, which are combined in a matrix with elements $C_{\nu,x}$, where ν enumerates the surrogate basis functions and x enumerates the measurement positions in the domain for which the observables are computed. We abbreviated the simulation data by the quantity $D_{\text{sim}} = \{A_s, Z_s\}$, where the matrix Z_s has the elements $(Z_s)_{i,x}$, which represent the observable $z^{(x)}$ at position

x corresponding to the model parameter vector $a_s^{(i)}$. The sought-for pdf follows from Bayes' theorem

$$p(C \mid D_{\text{sim}}) = p(C \mid Z_s, A_s) \propto p(Z_s \mid C, A_s)p(C \mid A_s) \propto p(Z_s \mid C, A_s).$$

The proportionality constant is not required in the ensuing considerations, and we have assumed an ignorant, uniform prior for the coefficients c, i.e., $p(C \mid \mathcal{I}) = const$. We note that this is also the transformation invariant Riemann prior (see Appendix B). However, any prior that is conjugate to the likelihood will retain analytical tractability. For the likelihood, we need the total misfit, which is given by

$$\chi^2 = \sum_{i=1}^{N_s} \sum_{x=1}^{N_x} \left((Z_s)_{i,x} - \sum_{v=1}^{N_p} (M_s)_{i,v}(C)_{v,x} \right)^2 = \sum_{i,x} \left(Z_s - M_s C \right)^2_{i,x}$$
$$= \text{tr}\left\{ \left(Z_s - M_s C \right)^T \left(Z_s - M_s C \right) \right\}, \tag{9}$$

with $(M_s)_{i,v} = \Phi_v(a_s^{(i)})$ and $N_{sx} = N_s \cdot N_x$. We assume a Gaussian type of likelihood, i.e.,

$$p(Z_s \mid C, A_s, \Delta) = \frac{\Delta^{-N_{sx}}}{Z'} \exp\left\{ -\frac{\chi^2}{2\Delta^2} \right\}. \tag{10}$$

with normalization Z'. We have mentioned the Δ-dependence of the normalization explicitly, while the rest of of the normalization is irrelevant in the present context. Usually, the misfit entering the likelihood comes from the noise of the data. In the present case, however, there is no noise (merely a tiny numerical error), but the surrogate model is presumably not an exact description of the simulation data and Δ covers the corresponding uncertainty. However, the uncertainty level Δ is not known and has to be marginalized over. Along with the appropriate Jeffreys' prior, $p(C, \Delta) = p(C)p(\Delta)$, $p(\Delta) = \frac{1}{\Delta}$, $p(C) = const.$ (see Appendix B), the integration over Δ yields

$$p(C \mid Z_s, A_s) = \frac{1}{Z} (\chi^2)^{-\frac{N_{sx}}{2}}. \tag{11}$$

with terms independent of C subsumed in the normalization Z. For computing the mean, variance, and evidence, we first complete the square in Equation (9) to get a quadratic form in C, which can then be integrated analytically (see Appendix A). The result is

$$\langle C \rangle_a = H_s^{-1} M_s^T Z_s, \qquad\qquad H_s = M_s^T M_s, \tag{12a}$$

$$\langle \Delta C_{vx} \Delta C_{v'x'} \rangle_a = \frac{\chi^2_{\min}}{(N_s - N_p)N_x - 2} \left(H_s^{-1} \right)_{v,v'} \delta_{xx'}, \qquad \chi^2_{\min} = \text{tr}\left\{ Z_s^T (\mathbf{1} - M_s H_s^{-1} M_s^T) Z_s \right\}. \tag{12b}$$

We argue that the prefactor of H_s^{-1} is the Bayesian estimate for Δ^2, the variance of the Gaussian in Equation (10). This reasoning is similar to [19]. Note that Z_s is a matrix of size $N_s \times N_x$, containing the data vectors of length N_s for each measurement site x. As shown in Appendix A, Equation (A6), the evidence for a particular set of surrogate models is computed as

$$p(\{z_{\text{sur}}^{(x)}\}_{x=1}^{N_x} \mid D_{\text{sim}}, \tilde{\mathcal{I}}) = Z = \Omega_{N_{px}} |H_s|^{-\frac{1}{2}} (\chi^2_{\min})^{-\frac{N_{sx} - N_{px}}{2}} \frac{\Gamma(\frac{N_{px}}{2})\Gamma(\frac{N_{sx} - N_{px}}{2})}{\Gamma(\frac{N_{sx}}{2})}, \tag{13}$$

where $\Omega_{N_{px}}$ is the solid angle in N_{px} dimensions. The evidence is the probability for a surrogate model given the data. Note that this quantity does not depend on $p(a \mid d_{\text{exp}}, \mathcal{I})$. This is reasonable because the analysis of the experimental data should be independent of the analysis of the simulation data. However, $p(a \mid d_{\text{exp}}, \mathcal{I})$ will typically be used for the experimental design of the simulation data acquisition. By comparing the evidence for different models, the user can choose a particular surrogate model or, if the results do not overwhelmingly suggest one single model, average the results for the surrogate analysis and the following uncertainty propagation over several plausible models. Note that the evidence is the pillar of a Bayesian procedure to select a surrogate model, and is distinct from the procedure of incorporating the trustworthiness or uncertainty of the surrogate in the subsequent uncertainty propagation.

2.3. Bayesian Uncertainty Propagation with Surrogate Models

Now that we have selected the surrogate model and determined the ingredients of Equation (7), we can determine the pdf for the observables in the light of the experimental data and the simulation results of the training set. The form in Equation (5) allows an easy evaluation of the mean value by using Equation (12) (see also Equation (A3)),

$$\left\langle z^{(x)} \right\rangle = \iint dV_a dV_C \, f(a \mid c^{(x)}) \, p(C \mid D_{\mathrm{sim}}, \tilde{\mathcal{I}}) \, p(a \mid d_{\mathrm{exp}}, \mathcal{I})$$
$$= \sum_v \int dV_a \Phi_v(a) \langle C_{vx} \rangle_a \, p(a \mid d_{\mathrm{exp}}, \mathcal{I}) \,. \tag{14}$$

Similarly, we obtain (see Equations (12) and (A8))

$$\left\langle z^{(x)} z^{(x')} \right\rangle = \iint dV_a dV_C \, f(a \mid c^{(x)}) \, f(a \mid c^{(x')}) \, p(C \mid D_{\mathrm{sim}}, \tilde{\mathcal{I}}) \, p(a \mid d_{\mathrm{exp}}, \mathcal{I})$$
$$= \sum_{vv'} \int dV_a \, \Phi_v(a) \, \Phi_{v'}(a) \, \langle C_{vx} C_{v'x'} \rangle_a \, p(a \mid d_{\mathrm{exp}}, \mathcal{I})$$
$$= \sum_{vv'} \int dV_a \, \Phi_v(a) \, \Phi_{v'}(a) \left(\langle C_{vx} \rangle_a \langle C_{v'x'} \rangle_a + \langle \Delta C_{vx} \Delta C_{v'x'} \rangle_a \right) \, p(a \mid d_{\mathrm{exp}}, \mathcal{I}) \,. \tag{15}$$

The covariance then follows from

$$\left\langle \Delta z^{(x)} \Delta z^{(x')} \right\rangle = \left\langle z^{(x)} z^{(x')} \right\rangle - \left\langle z^{(x)} \right\rangle \left\langle z^{(x')} \right\rangle \,. \tag{16}$$

If we neglected the uncertainty of the surrogate, i.e.,

$$p(C \mid D_{sim}) = \delta(C - \hat{C}) \,,$$
$$\hat{C} = \langle C \rangle_a \,,$$

then we retain the widely known special case of 'perfectly trustworthy' surrogates

$$\left\langle z^{(x)} z^{(x')} \right\rangle = \sum_{vv'} \int dV_a \, \Phi_v(a) \, \Phi_{v'}(a) \, \langle C_{vx} \rangle_a \langle C_{v'x'} \rangle_a \, p(a \mid d_{\mathrm{exp}}, \mathcal{I}) \,.$$

Thus, the first part in the integral of Equation (15) is the uncertainty of the observable due to experimental uncertainties and given the surrogate model, while the second term adds the uncertainty of the surrogate itself. The term $\langle \Delta C_{vx} \Delta C_{v'x'} \rangle_a$ is commonly neglected, but easily computed. This result also suggests a natural measure for the trustworthiness of the surrogate model, which is directly linked to the specific experiment:

$$\frac{\sum_{vv'} \int dV_a \, \Phi_v(a) \, \Phi_{v'}(a) \, \langle \Delta C_{vx} \Delta C_{v'x'} \rangle_a \, p(a \mid d_{\mathrm{exp}}, \mathcal{I})}{\sum_{vv'} \int dV_a \, \Phi_v(a) \, \Phi_{v'}(a) \, \langle C_{vx} \rangle_a \langle C_{v'x'} \rangle_a \, p(a \mid d_{\mathrm{exp}}, \mathcal{I})} < \epsilon \,. \tag{17}$$

If the surrogate uncertainties are, on average, smaller than the experimental uncertainties by a few orders of magnitude, e.g., $\epsilon = 10^{-3}$, then they may be neglected. However, ϵ is the user's choice. Note that this result does not spare the user to solve the foregoing surrogate model selection problem by e.g., computing evidence. This work only demonstrates how surrogate uncertainties can be incorporated and a practical rule when they could be neglected, given that the surrogate model has already been selected before.

3. Numerical Example

Here, we demonstrate an application where surrogate uncertainties were in part negligible and in part non-negligible. We apply our method to a computational fluid dynamics simulation of aortic hemodynamics, i.e., blood flow in an aorta resembled by the simplified geometry of an upside down umbrella stick. The simulation depends on a non-Newtonian viscosity model with four parameters $a = \{a_1, a_2, a_3, a_4\}$. The model was accompanied by viscosity measurements of human blood samples, thus determining $p(a \mid d_{\mathrm{exp}}, \mathcal{I})$. This posterior turned out to have a complex landscape that cannot be reasonably approximated by standard distributions. Particularly, strong correlations and multi-modality were observed, i.e., $p(a \mid d_{\mathrm{exp}}, \mathcal{I}) \neq \prod_i p(a_i \mid d_{\mathrm{exp}}, \mathcal{I})$. This means that vanilla Polynomial Chaos Expansions could not be applied without an undesirable transformation to conditionally independent variables. The posterior is described in detail in [20]. Based on $p(a \mid d_{\mathrm{exp}}, \mathcal{I})$, $N_s = 100$ parameter samples a_s were chosen and the simulation evaluated accordingly. The output, Z_s, was the absolute values of the wall shear stress that the blood flow exerts on the aortic wall, for $N_x = 10$ measurement probes at different locations, each for $N_t = 101$ time-instances equidistantly spaced over one cardiac cycle (ca. 1 s). Further details on the simulation are not relevant here but are documented [20]. A simulation time on the order of 150 CPU hours per sample suggested to use a surrogate for the inference. For the surrogate's basis functions, $\Phi_v(a)$, we found multivariate Legendre polynomials of up to order two sufficient. The numerical integrals were computed with Riemannian quadrature and convergence checked with successive grid refinement; however, stochastic integration would work just as well. A sketch example on how to implement this procedure computationally efficient via vectorisation in parameter space can be found at https://github.com/Sranf/BayesianSurrogate_sketch.git (1 January 2021).

In Figure 1, we compare the simulation uncertainty (including surrogate uncertainty) as computed with our Bayesian approach (Equation (15)) to the naive estimate for the simulation uncertainty (without surrogate uncertainty, i.e., neglecting $\langle \Delta C_{vx} \Delta C_{v'x'} \rangle_a$ in Equation (15). The surrogate uncertainties in the first half (left hand side) are relatively small, comprising only a few percent of the total uncertainty, and could possibly be neglected. In the second half (right-hand side), however, the surrogate uncertainties make up to $\sim$50% of the total uncertainty. This demonstrates that simulation uncertainties inferred via surrogate models can be severely underestimated if the surrogate uncertainties are neglected, and subsequently lead to overconfidence in the simulation model. In practice, one would acquire more data in order to reduce the surrogate uncertainties, e.g., more data at later time-instances in Figure 1 are particularly promising. This was limited here not only by the computational budget, but also the impracticality in that dynamic simulations require the full evaluation of all previous time instances where the surrogate is already reasonably accurate. Thus, the procedure of instead explicitly including the surrogate uncertainties here also has proven to be practical. A similar situation is to be expected for most transient simulations, as uncertainties will usually increase as time progresses.

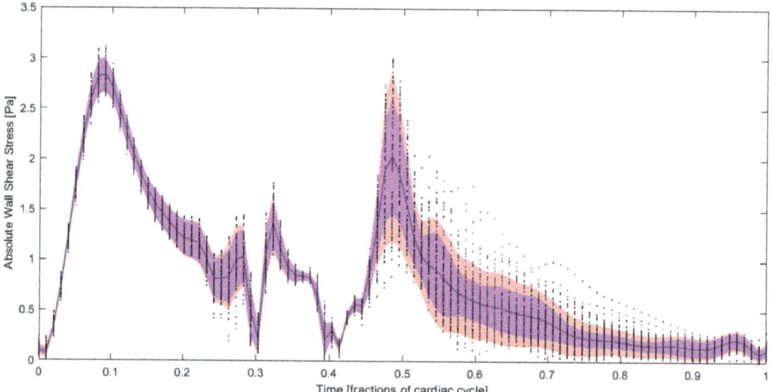

Figure 1. Simulation data (black dots) and simulation uncertainty (1σ) according to our Bayesian approach (red, including surrogate uncertainty) as well as the naive simulation uncertainty (blue, neglecting surrogate uncertainty). The black line is the surrogate mean.

4. Discussion

In this work, we have assumed a Gaussian likelihood for the simulation data, with unknown variance, for a surrogate that is linear in its parameters. Surrogates that are nonlinear in its parameters (e.g., neural networks) may promise higher capacity, however at the expense of losing analytical tractability of the surrogate uncertainty entirely. Other likelihood functions might be useful if further information is available, such as bounds on the observable (Gamma- or Beta-likelihood).

The result is a simple formula to incorporate surrogate uncertainties in the simulation uncertainties. This formula will be particularly useful if 'convergence' in the sense of finding the coefficients of e.g., a Polynomial Chaos Expansion is doubtful or not achievable due to a limit to the computational budget. The formula immediately suggests an intrinsic measure for the trustworthiness of the surrogate, distinct from commonly used ad hoc diagnostics. This measure is not to be confused with the evidence and should not be used for model selection because it would not preclude over-fitting, etc. It is merely a measure for the trustworthiness of the already selected surrogate.

Let us now explore the connections of this work to Polynomial Chaos Expansions (PCE) and Gaussian Process Regression (GPR). PCE is a special case of our generalized linear surrogate model, in that the basis functions of the surrogate are chosen such that

$$\int \Phi_v(a)\Phi_{v'}(a)p(a \mid d_{\exp}, \mathcal{I})dV_a := \delta_{v,v'} . \tag{18}$$

The double sum in Equation (15) then contracts to a single sum, and the diagonal of the term for the surrogate uncertainty, $\langle \Delta C_{vx}\Delta C_{v'x'}\rangle_a$, survives. This is expected, in that PCE is defined such that the basis functions are uncorrelated, but still the expansion coefficients must be uncertain to a finite degree, and this must carry over to the simulation uncertainty. A severe limitation of PCE is that it is rather difficult to find basis function sets $\{\Phi_v\}$ that fulfill Equation (18), depending on $p(a \mid d_{\exp}, \mathcal{I})$. For most practical purposes, one demands (i) conditional independence of the simulation parameters, i.e., $p(a \mid d_{\exp}, \mathcal{I}) = \prod_i p(a_i \mid d_{\exp}, \mathcal{I})$, as well as (ii) simple standard distributions for $p(a_i \mid d_{\exp}, \mathcal{I})$, in order to find a solution (usually a tensor-product) to Equation (18). Known albeit tedious workarounds are for (i) variable transformations and numerical orthonormalisation [18] and for (ii) PCE-constructions for arbitrary pdfs [17]. Note that also [17] demands (i) conditional independence of the simulation parameters. Note that our approach is not afflicted by above considerations, unless Equation (18) is specifically demanded. In the numerical example above, neither (i) nor (ii) were applicable. Finding a variable transformation in order to fulfill (i) or numerical construction of orthonormal basis functions can be difficult, and particularly inconvenient if sophisticated priors $p(a \mid \mathcal{I})$ are being used, e.g., Jeffreys' generalized prior. An interesting alternative would be to model the input dependencies with vine copulas [21] in order to overcome the limitations of PCE addressed here. Unfortunately, no obvious vine copula was found for the example presented here.

Gaussian Process Regression would correspond to a change in the prior for $z^{(x)}$ in Equation (6) as follows:

$$p\big(z^{(x)} \mid C, a, \theta, \tilde{\mathcal{I}}\big) = \mathcal{N}\Big(g(a \mid c^{(x)})\Big|K(\theta)\Big) . \tag{19}$$

where $\mathcal{N}$ denotes a normal distribution, and K is the prior's covariance matrix and defined by the parametrized covariance function k, $[K]_{ij} = k(a^{(i)}, a^{(j)} \mid \theta)$. This in turn would change $(Z_s - M_s C)^T (Z_s - M_s C) \to (Z_s - M_s C)^T K^{-1} (Z_s - M_s C)$ in Equation (9). By again completing the square and following the same procedure, the corresponding results for mean Equation (12a), variance Equation (12b), and evidence Equation (13) are then retained by a simple substitution of

$$H_s \to \tilde{H}_s , \qquad\qquad \tilde{H}_s = M_s^T K_s^{-1} M_s ,$$
$$\chi^2_{min} \to \tilde{\chi}^2_{min} , \qquad\qquad \tilde{\chi}^2_{min} = \mathrm{tr}\Big\{ Z_s^T (K_s^{-1} - K_s^{-1} M_s \tilde{H}_s M_s^T K_s^{-1}) Z_s \Big\} ,$$

where $[K_s]_{ij} = k(a_s^{(i)}, a_s^{(j)} \mid \theta)$ is the likelihood's covariance matrix evaluated at A_s for the data set Z_s at given θ. Equations (14) and (15) would preserve their form with the substitution

$$\Phi_v(a)\langle C \rangle_a \to \Phi_v(a)\langle C \rangle_{a,\theta} + K_*^T K_s^{-1} M_s \langle C \rangle_{a,\theta} ,$$
$$\langle \Delta C_{vx} \Delta C_{v'x'} \rangle_a \to \langle \Delta C_{vx} \Delta C_{v'x'} \rangle_{a,\theta} \Big(K - K_*^T K_s^{-1} K_* \Big) ,$$

where the subscript θ acknowledges that the right-hand side now depends on θ, and $[K_*]_{ij} = k(a^{(i)}, a_s^{(j)} \mid \theta)$ is the covariance between the training set A_s and the 'test set', i.e., the integration variable a. Note that the additionally introduced hyperparameters θ would require the choice of a prior for θ and marginalization wrt θ in Equation (7), and subsequently also Equations (12)–(15).

We now discuss the implications of $\tilde{\mathcal{I}}$ in contrast to the original background information $\mathcal{I}$. $\mathcal{I}$ contains, most importantly, that the observable z is uniquely determined by the simulation for a given set of input parameters a. A prerequisite here was that the simulation is converged. For example, for finite element simulations, this would be a given mesh-converged spatial discretization. The proposition $\tilde{\mathcal{I}}$ additionally assumes Equations (3) and (6), so that it can be used to replace Equation (2) by Equation (4). Formally, this means, to get from Equations (2)–(4), we replace $p(z \mid a, \mathcal{I}) \to p(z \mid a, c, \tilde{\mathcal{I}})$, where $p(z \mid a, \mathcal{I}) = \delta(z - z(a))$, $p(z \mid a, c, \tilde{\mathcal{I}}) = \delta(z - z_{sur}(a)) = \delta(z - g(a \mid c))$. For example, $\tilde{\mathcal{I}}$ contains in comparison to $\mathcal{I}$ the additional assumption that we can use the value for z as predicted/approximated by the surrogate model. It also means that we introduce additional, artificial, and usually unknown regression parameters c that need to be marginalized over. The additional uncertainty introduced by this approximation (i.e., the surrogate assumption) is encoded in $p(c|D_{sim}, \tilde{\mathcal{I}})$, and is correctly incorporated into the simulation observable uncertainties in Equation (15). What is important here is that $p(z \mid d_{exp}, D_{sim}, \tilde{\mathcal{I}}) \neq p(z \mid d_{exp}, D_{sim}, \mathcal{I})$ in general (the latter is computationally infeasible), but $p(z \mid d_{exp}, D_{sim}, \tilde{\mathcal{I}}) \approx p(z \mid d_{exp}, D_{sim}, \mathcal{I})$ if Equation (3) holds and $p(C \mid D_{sim}) \approx \delta(C - \hat{C})$, i.e., if the surrogate is indeed a good approximation and the posterior for the surrogate parameters is sharply peaked at $\hat{C}$. Very often, this posterior is not sharply peaked; then, we can just gather more data until it is, or, if that is not possible, we can at least avoid overconfidence induced by neglecting these uncertainties. A numerical example of where this is the case has been demonstrated above.

We have modeled spatial correlations by introducing a location index x, and assumed that the expansion coefficients at different sites, $c^{(x)}$ and $c^{(x')}$, are conditionally independent. This assumption is reasonable, in that the expansion coefficients are arbitrary mathematical constructs and no physically motivated model for their correlation is known. The spatial correlation, however, is retained in $z^{(x)}$, as was originally intended. More general models for spatial correlations can easily be implemented by substitution of $\delta_{xx'}$ with a spatial covariance matrix in Equation (12b). Note that this would require an additional marginalization wrt the (typically nonlinear) hyperparameters of the spatial covariance matrix. By introducing a compound index $\tilde{x} = (x, t)$ and substituting $x \to \tilde{x}$, we find a simple generalization to spatio-temporal correlations. This is equivalent to re-ordering spatial and temporal indices into a single sequence. While this procedure is convenient and requires only minor changes in the numerical implementation, it implicitly assumes conditional independence of spatial and temporal correlations. Analogous to above, general temporal correlations can be modeled by a substitution of $\delta_{\tilde{x}\tilde{x}'}$ in Equation (12b) with a temporal covariance matrix, again requiring an additional marginalization wrt the latter's hyperparameters.

5. Conclusions

We presented a Bayesian analysis of surrogate models and its associated uncertainty propagation problem in the context of uncertainty quantification of computer simulations. The assumptions were a generalized linear surrogate model (linear in its parameters, not the variable) and a Gaussian likelihood with unknown variance. Additionally, spatial and temporal correlations have been discussed. The result suggests a measure of trustworthiness of the surrogate by quantifying the ratio of the surrogate uncertainty to the total uncertainty, in contrast to commonly used heuristic diagnostics. The main result, however, is a rather simple rule to include surrogate uncertainties in the sought-for uncertainties of the simulation output. This is useful particularly for problems where the surrogate's trustworthiness is doubtful and cannot be improved. The connections to Polynomial Chaos Expansions and Gaussian Process Regression have been discussed. A numerical example demonstrated that simulation uncertainties can be significantly underestimated if surrogate uncertainties are neglected.

Funding: This work was funded by Graz University of Technology (TUG) through the LEAD Project "Mechanics, Modeling, and Simulation of Aortic Dissection" (biomechaorta.tugraz.at) and supported by GCCE: Graz Center of Computational Engineering.

Data Availability Statement: All information is contained in the manuscript. Code sketches are available at https://github.com/Sranf/BayesianSurrogate_sketch.git (1 January 2021).

Acknowledgments: The authors are grateful for useful comments from Ali Mohammad-Djafari.

Conflicts of Interest: The authors declare no conflict of interest.

Appendix A. Mathematical Proofs

Here, we want to determine norm, mean, and covariance of the marginalized Gaussian (Student-t distribution) in Equation (11), which is

$$p(C \mid Z_s, A_s) = \frac{1}{Z}(\chi^2)^{-\frac{N_s x}{2}},$$

$$\chi^2 = \mathrm{tr}\left\{(Z_s - M_s C)^T (Z_s - M_s C)\right\}. \tag{A1}$$

In order to perform the integration, we first complete the square to get a quadratic form in C, which can then be integrated analytically, i.e., we bring the misfit χ^2 into a form that elucidates the C-dependence

$$\chi^2 = \chi^2_{\min} + \text{tr}\left\{(C - \hat{C})^T H_s (C - \hat{C})\right\}, \qquad H_s = M_s^T M_s ,$$
$$\chi^2_{\min} = \text{tr}\left\{Z_s^T (\mathbf{1} - M_s H_s^{-1} M_s^T) Z_s\right\}, \qquad \hat{C} = H_s^{-1} M_s^T Z_s .$$

Now, the first moment is easily obtained. Along with the variable transformation under the integral

$$C \to \hat{C} + X \tag{A2}$$

we obtain

$$\langle C \rangle = \frac{1}{Z} \int dV_C \, C \left(\text{tr}\left\{(C - \hat{C})^T H_s (C - \hat{C})\right\} + \chi^2_{\min} \right)^{-\frac{N_{sx}}{2}}$$
$$= \hat{C} + \frac{1}{Z} \underbrace{\int dV_X \, X \left(X^T H_s X + \chi^2_{\min} \right)^{-\frac{N_{sx}}{2}}}_{=0} . \tag{A3}$$

where we have used the symmetry properties of the likelihood. Next, we transform the expression for normalization based on Equation (A2)

$$Z_{N_{sx}} = \int dV_X \left(\text{tr}\left\{X^T H_s X\right\} + \chi^2_{\min} \right)^{-\frac{N_{sx}}{2}} .$$

Now, we combine and reorder the double indices (ν, x) into a single index l, which turns the matrix X of dimension $N_p \times N_x$ into a vector x of dimension $N_{px} = N_p \cdot N_x$ and the matrix H of dimension $N_p \times N_p$ into a new block matrix H of dimension $N_{px} \times N_{sx}$ such that

$$[H]_{ll'} = [H_s]_{\nu,\nu'} \, \delta_{xx'} . \tag{A4}$$

In this representation, we have

$$Z_{N_{sx}} = \int dV_x \left(x^T H x + \chi^2_{\min} \right)^{-\frac{N_{sx}}{2}} = |H|^{-\frac{1}{2}} \int dV_y \left(y^T y + \chi^2_{\min} \right)^{-\frac{N_{sx}}{2}} , \tag{A5}$$

where we substituted $x \to H^{-\frac{1}{2}} y$. Next, we introduce hyper-spherical coordinates, which leads to

$$Z_{N_{sx}} = \Omega_{N_{px}} |H|^{-\frac{1}{2}} \int_0^\infty \frac{d\rho}{\rho} \, \rho^{N_{px}} (\rho^2 + \chi^2_{\min})^{-\frac{N_{sx}}{2}} ,$$

where $\Omega_{N_{px}}$ is the solid angle in N_{px} dimensions. Finally, based on the substitution $\rho = t \cdot \sqrt{\chi^2_{\min}}$, we recover an identity of the Beta-function, and we obtain

$$Z_{N_{sx}} = \Omega_{N_{px}} |H|^{-\frac{1}{2}} (\chi^2_{\min})^{-\frac{N_{sx} - N_{px}}{2}} \frac{\Gamma(\frac{N_{px}}{2}) \Gamma(\frac{N_{sx} - N_{px}}{2})}{\Gamma(\frac{N_{sx}}{2})} . \tag{A6}$$

This result is valid only for $N_{sx} > N_{px}$, which is fulfilled in the present application. For future use, we rewrite this as

$$Z_{N_{sx}} = Z_{(N_{sx}-2)} \cdot \left(\chi^2_{\min}\right)^{-1} \cdot \frac{N_{sx} - N_{px} - 2}{N_{sx} - 2} .$$

(A7)

Finally, we calculate the covariance, based also on the compound index $l = (\nu, x)$, and by using the variable transformation in Equation (A2):

$$
\begin{aligned}
\langle \Delta C_l \Delta C_{l'} \rangle &= \frac{1}{Z_{N_{sx}}} \int dV_x \, x_l x_{l'} \left(x^T H x + \chi^2_{\min} \right)^{-\frac{N_{sx}}{2}} , \\
&= -\frac{2}{N_{sx} - 2} \cdot \frac{1}{Z_{N_{sx}}} \cdot \frac{\partial}{\partial H_{l,l'}} \int dV_x \left(x^T H x + \chi^2_{\min} \right)^{-\frac{N_{sx}}{2}+1} , \\
&= -\frac{2}{N_{sx} - 2} \frac{\chi^2_{\min} \cdot (N_{sx} - 2)}{Z_{(N_{sx}-2)} \cdot (N_{sx} - N_{px} - 2)} \frac{\partial}{\partial H_{ll'}} Z_{(N_{sx}-2)} , \\
&= -\frac{2\chi^2_{\min}}{(N_{sx} - N_{px} - 2)} \frac{\partial}{\partial H_{ll'}} \ln(Z_{(N_{sx}-2)}) , \\
&= -\frac{2\chi^2_{\min}}{(N_{sx} - N_{px} - 2)} \underbrace{\frac{\partial}{\partial H_{ll'}} \ln(|H|^{-\frac{1}{2}})}_{=-\frac{1}{2}\left(H^{-1}\right)_{ll'}} .
\end{aligned}
$$

(A8)

In the last step, we have used that H is a symmetric matrix. This is a very reasonable result because, if the variance Δ^2 in the Gaussian in Equation (10) would be known, then the covariance is $\Delta^2 H^{-1}$. Consequently, the prefactor represents the Bayesian estimate for the variance Δ^2 based on the data. Now, we go back to the original meaning of the compound index Equation (A4), i.e., $\left(H^{-1}\right)_{ll'} \to (H_s^{-1})_{\nu\nu'}\delta_{xx'}$, and obtain the final result.

Appendix B. The Transformation Invariant Prior for the Surrogate Coefficients

Bayesian probability theory allows for rigorously and consistently incorporating any prior knowledge we have about the experiment before taking a look at the data. This knowledge shall be elicited here. Our inference must not depend on the exact parametrization. e.g., if we re-parametrize the surrogate, re-label, or re-order the surrogate parameters, the surrogate still should describe the same simulation. This is reasonable because the surrogate is a purely mathematical, auxiliary construct. This rescaling-invariance is ensured by Jeffreys' generalized prior and is given by the Riemann metric R (or the determinant of the Fisher information matrix) [16]

$$p(C) = \frac{1}{Z}|\det(R)|^{1/2} \quad \text{with} \quad R_{ij} = \int p(Z_s \mid C)\frac{\partial^2}{\partial C_i \partial C_j} \ln\left(p(Z_s \mid C)\right) dV_{Z_s} .$$

(A9)

with multi-indices $i, j = (\nu, x)$. With the likelihood and the generalized surrogate model defined in the manuscript, the result is

$$
\begin{aligned}
R_{ij} &\propto \sum_{k=1}^{N_s} \frac{\partial g(a_s^{(k)} \mid C)}{\partial C_i} \frac{\partial g(a_s^{(k)} \mid C)}{\partial C_j} \\
&= \sum_{k=1}^{N_s} \Phi_i(a_s^{(k)})\Phi_j(a_s^{(k)}) \\
&= const.
\end{aligned}
$$

(A10)

This prior is independent of C, i.e., a constant.

References

1. Xiu, D.; Karniadakis, G.E. The Wiener-Askey polynomial chaos for stochastic differential equations. *SIAM J. Sci. Comput.* **2005**, *27*, 1118–1139. [CrossRef]
2. O'Hagan, A. Polynomial Chaos: A Tutorial and Critique from a Statistician's Perspective. 2013. Available online: http://tonyohagan.co.uk/academic/pdf/Polynomial-chaos.pdf (accessed on 25 June 2019).
3. Crestaux, T.; Le Maître, O.P.; Martinez, J.-M. Polynomial chaos expansion for sensitivity analysis. *Reliab. Eng. Syst. Saf.* **2009**, *94*, 1161–1172. [CrossRef]
4. O'Hagan, A. Curve Fitting and Optimal Design for Prediction. *J. R. Stat. Soc. Ser. B* **1978**, *40*, 1–42. [CrossRef]
5. Rasmussen, C.E.; Williams, C.K. *Gaussian Processes for Machine Learning*; The MIT Press: Cambridge, MA, USA, 2006.
6. Sraj, I.; Le Maître, O.P.; Knio, O.M.; Hoteit, I. Coordinate transformation and Polynomial Chaos for the Bayesian inference of a Gaussian process with parametrized prior covariance function. *Comput. Methods Appl. Mech. Eng.* **2016**, *298*, 205–228. [CrossRef]
7. O'Hagan, A.; Kennedy, M.C.; Oakley, J.E. Uncertainty analysis and other inference tools for complex computer codes. *Bayesian Stat.* **1999**, *6*, 503–524.
8. Kennedy, M.C.; O'Hagan, A. Predicting the output from a complex computer code when fast approximations are available. *Biometrika* **2000**, *87*, 1–13. [CrossRef]
9. Arnst, M.; Ghanem, R.G.; Soize, C. Identification of Bayesian posteriors for coefficients of chaos expansions. *J. Comput. Phys.* **2010**, *229*, 3134–3154. [CrossRef]
10. Madankan, R.; Singla, P.; Singh, T.; Scott, P.D. Polynomial-chaos-based Bayesian approach for state and parameter estimations. *J. Guid. Control Dyn.* **2013**, *36*, 1058–1074. [CrossRef]
11. Karagiannis, G.; Lin, G. Selection of polynomial chaos bases via Bayesian model uncertainty methods with applications to sparse approximation of PDEs with stochastic inputs. *J. Comput. Phys.* **2014**, *259*, 114–134. [CrossRef]
12. Lu, F.; Morzfeld, M.; Tu, X.; Chorin, A.J. Limitations of polynomial chaos expansions in the Bayesian solution of inverse problems. *J. Comput. Phys.* **2015**, *282*, 138–147. [CrossRef]
13. Hwai, M.; Tan, Y. Sequential Bayesian Polynomial Chaos Model Selection for Estimation of Sensitivity Indices. *SIAM/ASA J. Uncertain. Quantif.* **2015**, *3*, 146–168. [CrossRef]
14. Ghanem, R.G.; Owhadi, H.; Higdon, D. *Handbook of Uncertainty Quantification*; Springer: New York, NY, USA, 2017; [CrossRef]
15. O'Hagan, A. Bayesian analysis of computer code outputs: A tutorial. *Reliab. Eng. Syst. Saf.* **2006**, *91*, 1290–1300. [CrossRef]
16. von der Linden, W.; Dose, V.; von Toussaint, U. *Bayesian Probability Theory: Applications in the Physical Sciences*, 1st ed.; Cambridge University Press: Cambridge, UK, 2014; [CrossRef]
17. Oladyshkin, S.; Nowak, W. Data-driven uncertainty quantification using the arbitrary polynomial chaos expansion. *Reliab. Eng. Syst. Saf.* **2012**, *106*, 179–190. [CrossRef]
18. Jakeman, J.D.; Franzelin, F.; Narayan, A.; Eldred, M.; Plfüger, D. Polynomial chaos expansions for dependent random variables. *Comput. Methods Appl. Mech. Eng.* **2019**, *351*, 643–666. [CrossRef]
19. von der Linden, W.; Preuss, R.; Hanke, W. Consistent Application of Maximum Entropy to Quantum-Monte-Carlo Data. *J. Physics: Condens. Matter* **1996**, *8*, 1–13. [CrossRef]
20. Ranftl, S.; Müller, T.; Windberger, U.; von der Linden, W.; Brenn, G. *Data and Codes for 'A Bayesian Approach to Blood Rheological Uncertainties in Aortic Hemodynamcis'*; Zenodo Digital Repository: Genève, Switzerland, 2021; [CrossRef]
21. Torre, E.; Marelli, S.; Embrechts, P.; Sudret, B. A general framework for data-driven uncertainty quantification under complex input dependencies using vine copulas. *Probabilistic Eng. Mech.* **2019**, *55*, 1–16. [CrossRef]

Proceeding Paper

Orbit Classification and Sensitivity Analysis in Dynamical Systems Using Surrogate Models [†]

Katharina Rath [1,2,*], Christopher G. Albert [2], Bernd Bischl [1] and Udo von Toussaint [2]

1   Department of Statistics, Ludwig-Maximilians-Universität München, 80333 Munich, Germany; bernd.bischl@stat.uni-muenchen.de
2   Max-Planck-Institut für Plasmaphysik, 85748 Garching, Germany; albert@alumni.tugraz.at (C.G.A.); udo.v.toussaint@ipp.mpg.de (U.v.T.)
*   Correspondence: katharina.rath@ipp.mpg.de
†   Presented at the 40th International Workshop on Bayesian Inference and Maximum Entropy Methods in Science and Engineering, online, 4–9 July 2021.

Abstract: Dynamics of many classical physics systems are described in terms of Hamilton's equations. Commonly, initial conditions are only imperfectly known. The associated volume in phase space is preserved over time due to the symplecticity of the Hamiltonian flow. Here we study the propagation of uncertain initial conditions through dynamical systems using symplectic surrogate models of Hamiltonian flow maps. This allows fast sensitivity analysis with respect to the distribution of initial conditions and an estimation of local Lyapunov exponents (LLE) that give insight into local predictability of a dynamical system. In Hamiltonian systems, LLEs permit a distinction between regular and chaotic orbits. Combined with Bayesian methods we provide a statistical analysis of local stability and sensitivity in phase space for Hamiltonian systems. The intended application is the early classification of regular and chaotic orbits of fusion alpha particles in stellarator reactors. The degree of stochastization during a given time period is used as an estimate for the probability that orbits of a specific region in phase space are lost at the plasma boundary. Thus, the approach offers a promising way to accelerate the computation of fusion alpha particle losses.

Keywords: Gaussian process regression; surrogate model; Lyapunov exponent; sensitivity analysis; Hamiltonian systems

Citation: Rath, K.; Albert, C.G.; Bischl, B.; von Toussaint, U. Orbit Classification and Sensitivity Analysis in Dynamical Systems Using Surrogate Models. *Phys. Sci. Forum* **2021**, 3, 5. https://doi.org/10.3390/psf2021003005

Published: 5 November 2021

Publisher's Note: MDPI stays neutral with regard to jurisdictional claims in published maps and institutional affiliations.

1. Introduction

Hamilton's equations describe the dynamics of many classical physics systems such as classical mechanics, plasma physics or electrodynamics. In most of these cases, chaos plays an important role [1]. One fundamental question in analyzing these chaotic Hamiltonian systems is the distinction between regular and chaotic regions in phase space. A commonly used tool are Poincaré maps, which connect subsequent intersections of orbits with a lower-dimensional subspace, called Poincaré section. For example, in a planetary system one could record a section each time the planet has made a turn around the Sun. The resulting pattern of intersection points on this subspace allow insight into the dynamics of the underlying system: regular orbits stay bound to a closed hyper-surface and do not leave the confinement volume, whereas chaotic orbits might spread over the whole phase space. This is related to the breaking of KAM (Kolmogorov-Arnold-Moser) surfaces that form barriers for motion in phase space [2]. The classification of regular versus chaotic orbits is performed, e.g., via box-counting [3] or by calculating the spectrum of Lyapunov exponents [4–6]. Lyapunov exponents measure the asymptotic average exponential rate of divergence of nearby orbits in phase space over infinite time and are therefore invariants of the dynamical system. When considering only finite time, the obtained *local* Lyapunov exponents (LLEs) for a specific starting position depend on the position in phase space and give insight into the local predictability of the dynamical system of interest [7–10].

Poincaré maps are in most cases inefficient to compute as their computation involves numerical integration of Hamilton's equations even though only intersections with the surface of interest are recorded. When using a surrogate model to interpolate the Poincaré map, the symplectic structure of phase space arising from the description in terms of the Hamiltonian description has to be preserved to obtain long-term stability and conservation of invariants of motion, e.g., volume preservation. Additional information on Hamiltonian systems and symplecticity can be found in [2,11]. Here, we use a structure-preserving Gaussian process surrogate model (SympGPR) that interpolates directly between Poincaré sections and thus avoids unnecessary computation while achieving similar accuracy as standard numerical integration schemes [12].

In the present work, we investigate how the symplectic surrogate model [12] can be used for early classification of chaotic versus regular trajectories based on the calculation of LLEs. The latter are calculated using the Jacobian that is directly available from the surrogate model [13]. As LLEs also depend on time, we study their distribution on various time scales to estimate the needed number of mapping iterations. We combine the orbit classification with a sensitivity analysis based on variance decomposition [14–16] to evaluate the influence of uncertain initial conditions in different regions of phase space. The analysis is carried out on the well-known standard map [17] that is well suited for validation purposes as a closed form expression for the Poincaré maps is available. This, however, does not influence the performance of the surrogate model that is applicable also in cases where such a closed form doesn't exist [12].

The intended application is the early classification of regular and chaotic orbits of fusion alpha particles in stellarator reactors [3]. While regular particles can be expected to remain confined indefinitely, only chaotic orbits have to be traced to the end. This offers a promising way to accelerate loss computations for stellarator optimization.

2. Methods

2.1. Hamiltonian Systems

A f−dimensional system (with $2f$−dimensional phase space) described by its Hamiltonian $H(\boldsymbol{q}, \boldsymbol{p}, t)$ depending on f generalized coordinates $\boldsymbol{q}$ and f generalized momenta $\boldsymbol{p}$ satisfies Hamilton's canonical equations of motion,

$$\dot{\boldsymbol{q}}(t) = \frac{d\boldsymbol{q}(t)}{dt} = \nabla_{\boldsymbol{p}} H(\boldsymbol{q}(t), \boldsymbol{p}(t)), \qquad \dot{\boldsymbol{p}}(t) = \frac{d\boldsymbol{p}(t)}{dt} = -\nabla_{\boldsymbol{q}} H(\boldsymbol{q}(t), \boldsymbol{p}(t)), \qquad (1)$$

which represent the time evolution as integral curves of the Hamiltonian vector field.

Here, we consider the standard map [17] that is a well-studied model to investigate chaos in Hamiltonian systems. Each mapping step corresponds to one Poincaré map of a periodically kicked rotator:

$$p_{n+1} = (p_n + K\sin(q_n)) \bmod 2\pi, \qquad q_{n+1} = (q_n + p_{n+1}) \bmod 2\pi, \qquad (2)$$

where K is the stochasticity parameter corresponding to the intensity of the perturbation. The standard map is an area-preserving map with $\det J = 1$, where J is its Jacobian:

$$J = \begin{pmatrix} \frac{\partial q_{n+1}}{\partial q_n} & \frac{\partial q_{n+1}}{\partial p_n} \\ \frac{\partial p_{n+1}}{\partial q_n} & \frac{\partial p_{n+1}}{\partial p_n} \end{pmatrix} = \begin{pmatrix} 1 + K\cos(q_n) & 1 \\ K\cos(q_n) & 1 \end{pmatrix} \qquad (3)$$

2.2. Symplectic Gaussian Process Emulation

A Gaussian process (GP) [18] is a collection of random variables, any finite number of which have a joint Gaussian distribution. A GP is fully specified by its mean $\boldsymbol{m}(\boldsymbol{x})$ and kernel or covariance function $K(\boldsymbol{x}, \boldsymbol{x}')$ and is denoted as

$$\boldsymbol{f}(\boldsymbol{x}) \sim \mathcal{GP}(\boldsymbol{m}(\boldsymbol{x}), K(\boldsymbol{x}, \boldsymbol{x}')), \qquad (4)$$

for input data points $x \in \mathbb{R}^d$. Here, we allow vector-valued functions $f(x) \in \mathbb{R}^D$ [19]. The covariance function is a positive semidefinite matrix-valued function, whose entries $(K(x, x'))_{ij}$ express the covariance between the output dimensions i and j of $f(x)$.

For regression, we rely on observed function values $Y \in \mathbb{R}^{D \times N}$ with entries $y = f(x) + \epsilon$. These observations may contain local Gaussian noise ϵ, i.e., the noise is independent at different positions x but may be correlated between components y. The input variables are aggregated in the $d \times N$ design matrix X, where N is the number of training data points. The posterior distribution, after taking training data points into account, is still a GP with updated mean $F_* \equiv \mathbb{E}(F(X_*))$ and covariance function allowing to make predictions for test data X_*:

$$F_* = K(X_*, X)(K(X, X) + \Sigma_n)^{-1} Y, \tag{5}$$

$$\mathrm{cov}(F_*) = K(X_*, X_*) - K(X_*, X)(K(X, X) + \Sigma_n)^{-1} K(X, X_*), \tag{6}$$

where $\Sigma_n \in \mathbb{R}^{ND \times ND}$ is the covariance matrix of the multivariate output noise for each training data point. Here we use the shorthand notation $K(X, X)$ for the block matrix assembled over the output dimension D in addition to the number of input points as in a single-output GP with a scalar covariance function $k(x, x')$ that expresses the covariance of different input data points x and x'. The kernel parameters are estimated given the input data by minimizing the negative log-likelihood [18].

To construct a GP emulator that interpolates symplectic maps for Hamiltonian systems, symplectic Gaussian process regression (SympGPR) was presented in [12] where the generating function $F(q, P)$ and its gradients are interpolated using a multi-output GP with derivative observations [20,21]. The generating function links old coordinates $(q, p) = (q_n, p_n)$ to new coordinates $(Q, P) = (q_{n+1}, p_{n+1})$ (e.g., after one iteration of the standard map Equation (2)) via a canonical transformation such that the symplectic property of phase space is preserved. Thus, input data points consist of pairs (q, P). Then, the covariance matrix contains the Hessian of an original scalar covariance function $k(q, P, q', P')$ as the lower block matrix $L(q, P, q', P')$ (denoted with the red box):

$$K(q, P, q', P') = \begin{pmatrix} k & \partial_{q'} k & \partial_{p'} k \\ \partial_q k & \boxed{\partial_{qq'} k} & \boxed{\partial_{qp'} k} \\ \partial_p k & \boxed{\partial_{Pq'} k} & \boxed{\partial_{pp'} k} \end{pmatrix}. \tag{7}$$

Using the algorithm for the (semi-)implicit symplectic GP map as presented in [12], once the SympGPR model is trained and the covariance matrix calculated, the model is used to predict subsequent time steps or Poincaré maps for arbitrary initial conditions.

For the estimation of the Jacobian (Equation (3)) from the SympGPR, the Hessian of the generating function $F(q, P)$ has to be inferred from the training data. Thus, the covariance matrix is extended with a block matrix C containing third derivatives of $k(q, P, q', P')$:

$$C = \begin{pmatrix} \partial_{q,q',q} k & \partial_{q,P',q} k & \partial_{P,q',q} k & \partial_{P,P',q} k \\ \partial_{q,q',p} k & \partial_{q,P',p} k & \partial_{P,q',p} k & \partial_{P,P',p} k \end{pmatrix}. \tag{8}$$

The mean of the posterior distribution of the desired Hessian of the generating function $F(q, P)$ is inferred via

$$\nabla^2 F = (\partial^2_{qq} F, \partial^2_{qP} F, \partial^2_{Pq} F, \partial^2_{PP} F)^\top = C L^{-1} Y. \tag{9}$$

As we have a dependence on mixed coordinates $Q(\bar{q}(q, p), P(q, p))$ and $P(Q(q, p), \bar{p}(q, p))$, where we used $\bar{q}(q, p) = q$ and $\bar{p}(q, p) = p$ to correctly carry out the inner derivatives, the needed elements for the Jacobian can be calculated employing the chain rule. The Jacobian is then given as the solution of the well-determined linear set of equations:

$$\frac{\partial Q}{\partial q} = \frac{\partial Q}{\partial \bar{q}}\frac{\partial \bar{q}}{\partial q} + \frac{\partial Q}{\partial P}\frac{\partial P}{\partial q}, \qquad \frac{\partial Q}{\partial p} = \frac{\partial Q}{\partial \bar{q}}\frac{\partial \bar{q}}{\partial p} + \frac{\partial Q}{\partial P}\frac{\partial P}{\partial p}, \tag{10}$$

$$\frac{\partial P}{\partial q} = \frac{\partial P}{\partial Q}\frac{\partial Q}{\partial q} + \frac{\partial P}{\partial \bar{p}}\frac{\partial \bar{p}}{\partial q}, \qquad \frac{\partial P}{\partial p} = \frac{\partial P}{\partial Q}\frac{\partial Q}{\partial p} + \frac{\partial P}{\partial \bar{p}}\frac{\partial \bar{p}}{\partial p}, \tag{11}$$

where we use the following correspondence to determine all factors of the SOEs:

$$\begin{pmatrix} \frac{\partial Q}{\partial \bar{q}} & \frac{\partial Q}{\partial P} \\ \frac{\partial \bar{p}}{\partial \bar{q}} & \frac{\partial \bar{p}}{\partial P} \end{pmatrix} = \begin{pmatrix} \frac{\partial \bar{q}}{\partial Q} & \frac{\partial P}{\partial Q} \\ \frac{\partial \bar{q}}{\partial \bar{p}} & \frac{\partial P}{\partial \bar{p}} \end{pmatrix}^{\top} = \begin{pmatrix} 1 + \frac{\partial^2 F}{\partial q \partial P} & -\frac{\partial^2 F}{\partial P \partial P} \\ -\frac{\partial^2 F}{\partial q \partial q} & 1 + \frac{\partial^2 F}{\partial P \partial q} \end{pmatrix}. \tag{12}$$

2.3. Sensitivity Analysis

Variance-based sensitivity analysis decomposes the variance of the model output into portions associated with uncertainty in the model inputs or initial conditions [14,15]. Assuming independent input variables X_i, $i = 1, ..., d$, the functional analysis of variance (ANOVA) allows a decomposition of the scalar model output Y from which the decomposition of the variance can be deduced:

$$V[Y] = \sum_{i=1}^{d} V_i + \sum_{1 \leq i < j \leq d} V_{ij} + ... + V_{1,2,...,d} \tag{13}$$

The first term describes the variation in variance only due to changes in single variables X_i, whereas higher-order interactions are depicted in the contributions of the interaction terms. From this, first-order Sobol' indices S_i are defined as the corresponding fraction of the total variance, whereas *total* Sobol' indices S_{T_i} also take the influence of X_i interacting with other input variables into account [14,15]:

$$S_i = \frac{V_i}{\text{Var}(Y)}, \qquad S_{T_i} = \frac{E_{\mathbf{X}_{\sim i}}(\text{Var}_{X_i}(Y|\mathbf{X}_{\sim i}))}{\text{Var}(Y)} \tag{14}$$

Several methods for efficiently calculating Sobol' indices have been presented, e.g., MC sampling [14,16] or direct estimation from surrogate models [22,23]. Here, we use the MC sampling strategy presented in [16] using two sampling matrices A, B and a combination of both $A_B^{(i)}$, where all columns are from A except the i-th column which is from B:

$$S_i \text{Var}(Y) = \frac{1}{N} \sum_{i=1}^{N} f(\mathbf{B})_j (f(A_B^{(i)})_j - f(A)_j), \qquad S_{T_i} \text{Var}(Y) = \frac{1}{2N} \sum_{i=1}^{N} (f(A)_j - f(A_B^{(i)})_j)^2, \tag{15}$$

where f denotes the model to be evaluated.

2.4. Local Lyapunov Exponents

For a dynamical system in $\mathbb{R}^D$, D Lyapunov characteristic exponents λ_n give the exponential separation of trajectories with initial conditions $z(0) = (q(0), p(0))$ of a dynamical system with perturbation δz over time:

$$|\delta z(T)| = \mathcal{J}_{z(T)}^{(T)} \delta z(0) \approx e^{T\lambda}|\delta z(0)|, \tag{16}$$

where $\mathcal{J}_{z(T)}^{(T)}$ is a time-ordered product of Jacobians $\mathcal{J}_{z(T-1)}\mathcal{J}_{z(T-2)}\cdots\mathcal{J}_{z(1)}\mathcal{J}_{z(0)}$ [4]. The Lyapunov exponents are then given as the logarithm of the eigenvalues of the positive and symmetric matrix.

$$\Lambda = \lim_{T \to \infty} [\mathcal{J}_{z(T)}^{(T)\top} \mathcal{J}_{z(T)}^{(T)}]^{1/(2T)}, \tag{17}$$

where $\top$ denotes the transpose of $\mathcal{J}_{z(T)}^{(T)}$.

For a D-dimensional system, there exist D Lyapunov exponents λ_n giving the rate of growth of a D-volume element with $\lambda_1 + ... + \lambda_D$ corresponding to the rate of growth of the determinant of the Jacobian $\det(\mathcal{J}_{z(T)}^{(T)})$. From this follows that for a Hamiltonian system with a symplectic (e.g., volume-preserving) phase space structure, Lyapunov exponents exist in additive inverse pairs as the determinant of the Jacobian is constant, $\lambda_1 + ... + \lambda_D = 0$. In the dynamical system of the standard map with $D = 2$ considered here, the Lyapunov exponents allow a distinction between regular and chaotic motion. If the Lyapunov exponents $\lambda_1 = -\lambda_2 > 0$, neighboring orbits separate exponentially which corresponds to a chaotic region. In contrast, when $\lambda_1 = -\lambda_2 \approx 0$ the motion is regular [1].

As the product of Jacobians is ill-conditioned for large values of T, several algorithms have been proposed to calculate the spectrum of Lyapunov exponents [13]. Here, we determine *local* Lyapunov exponents (LLE) that determine the predictability of an orbit of the system at a specific phase point for finite time. In contrast to *global* Lyapunov exponents they depend on T and on the position in phase space z. We use recurrent Gram-Schmidt orthonormalization procedure through QR decomposition [5,6,24], where we follow the evolution of D initially orthonormal deviation vectors w_0^n. The Jacobian is decomposed into $\mathcal{J}_{z(0)} = Q^{(1)}R^{(1)}$, where $Q^{(1)}$ is an orthogonal matrix and $R^{(1)}$ is an upper triangular matrix yielding a new set of orthonormal vectors w_i. At the next mapping iteration, the matrix product $\mathcal{J}_{z(1)}Q^{(1)}$ is again decomposed. This procedure is repeated T times to arrive at $\mathcal{J}_{z(t)}^{(T)} = Q^{(T)}R^{(T)}R^{(T-1)}...R^{(0)}$. The Lyapunov exponents are then estimated from the diagonal elements of $R^{(t)}$

$$\lambda_n = \frac{1}{T} \sum_{t=1}^{T} \ln R_{nn}^{(t)}. \tag{18}$$

3. Results and Discussion

In the following we apply an implicit SympGPR model with a product kernel [12]. Due to the periodic topology of the standard map we use a periodic kernel function to construct the covariance matrix in Equation (7) with periodicity 2π in q, whereas a squared exponential kernel is used in P:

$$k(q, q_i, P, P_i) = \sigma_f^2 \exp\left(-\frac{\sin^2((q - q_i)/2)}{2l_q^2}\right) \exp\left(-\frac{(P - P_i)^2}{2l_P^2}\right). \tag{19}$$

Here σ_f^2 specifies the amplitude of the fit and is set in accordance with the observations to $2\max(|Y|)^2$, where Y corresponds to the change in coordinates. The hyperparameters l_q, l_P are set to their maximum likelihood value by minimizing the negative log-likelihood given the input data using the L-BFGS-B routine implemented in Python [18]. The noise in observations is set to $\sigma_n^2 = 10^{-16}$. 30 initial data points are sampled from a Halton sequence to ensure good coverage of the training region in the range $[0, 2\pi] \times [0, 2\pi]$ and Equation (2) is evaluated once to obtain the corresponding final data points. Each pair of initial and final conditions constitutes one sample of the training data set. Once the model is trained, it is used to predict subsequent mapping steps for arbitrary initial conditions and to infer the corresponding Jacobians for the calculation of the local Lyapunov exponents. Here, we consider two test cases of the standard map with different values of the stochasticity parameter $K = 0.9$ and $K = 2.0$ (Equation (2)). For each of the test cases, a surrogate model is trained. While in the first case the last KAM surface is not yet broken and therefore the region of stochasticity is still confined in phase space, in the latter case the chaotic region covers a much larger portion of phase space. However, there still exist islands of stability with regular orbits [2]. For $K = 0.9$ the mean squared error (MSE) for the training data is 1.4×10^{-6}, whereas the test MSE after one mapping application is found to be 2.4×10^{-6}. A similar quality of the surrogate model is reached for $K = 2.0$, where the training MSE is 1.6×10^{-7} and the test MSE 2.4×10^{-7}.

3.1. Local Lyapunov Exponents and Orbit Classification

For the evaluation of the distribution of the local Lyapunov exponents with respect to the number of mapping iterations T and phase space position $z = (q, p)$, 1000 points are sampled from each orbit under investigation. In the following, we only consider the maximum local Lyapunov exponent as it determines the predictability of the system. For each of the 1000 points, the LLEs are calculated using Equation (18), where the needed Jacobians are given by the surrogate model by evaluating Equation (9) and solving Equation (11).

Figure 1 shows the distributions for $K = 2.0$, $T = 50$, $T = 100$ and $T = 1000$ for two different initial conditions resulting in a regular and a chaotic orbit. In the regular case the distribution exhibits a sharp peak and with increasing T moves closer to 0. This bias due to the finite number of mapping iterations decreases with $\mathcal{O}(1/T)$ as shown in Figure 2 [25]. For the chaotic orbit, the distribution looks smooth and its median is clearly >0 as expected. For a smaller value of $K = 0.9$ the dynamics in phase space exhibit larger variety with regular, chaotic and also weakly chaotic orbits that remain confined in a small stochastic layer around hyperbolic points. Hence, the transition between regular, weakly chaotic and chaotic orbits is continuous due to the larger variety in phase space. For fewer mapping iterations, possible values of λ are overlapping, thus preventing a clear distinction between confined chaotic and chaotic orbits.

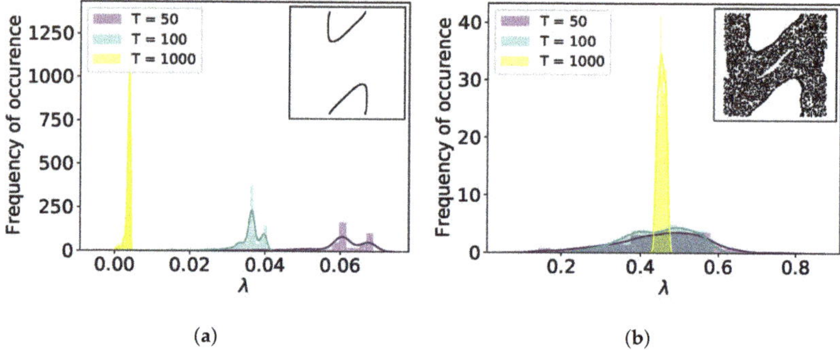

(a)   (b)

Figure 1. Distribution of local Lyapunov exponents for a (**a**) regular orbit $(q, p) = (1.96, 4.91)$ and (**b**) chaotic orbit $(q, p) = (0.39, 2.85)$ in the standard map with $K = 2.0$

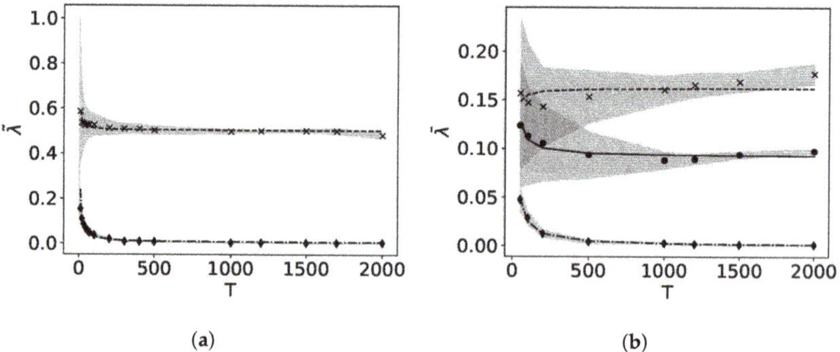

(a)   (b)

Figure 2. Rate of convergence of the block bias due to finite number of mapping iterations for (**a**) $K = 2.0$ with a regular orbit $(q, p) = (1.96, 4.91)$ (diamond) and a chaotic orbit $(q, p) = (0.39, 2.85)$ (x) and (**b**) $K = 0.9$ with a regular orbit $(q, p) = (1.76, 0.33)$ (diamond), a confined chaotic orbit $(q, p) = (0.02, 2.54)$ (circle) and a chaotic orbit $(q, p) = (0.2, 5.6)$ (x). The graphs show $\tilde{\lambda}_T$, the median of λ_T for each T, with $\tilde{\lambda}_T = \lambda + c/T$ fitted by linear regression of $T\tilde{\lambda}_T$ on T. The gray areas correspond to the standard deviation for 1000 test points.

When considering the whole phase space with 200 orbits with initial conditions sampled from a Halton sequence in the range $[0, \pi] \times [0, 2\pi]$, already $T = 50$ mapping iterations provide insight in the predictability of the standard map (Figure 3). If for a region in phase space the obtained LLE is positive, the predictability in this region is restricted as the instability there is relatively large. If, however, the LLE is close to zero, we can conclude that this region in phase space is governed by regular motion and is therefore highly predictable. For $K = 2.0$ the orbits constituting the chaotic sea have large positive LLEs, whereas islands of stability built by regular orbits show LLEs close to 0. A similar behavior can be observed for $K = 0.9$, where again regions around stable elliptic points feature $\lambda \approx 0$ while stochastic regions exhibit a varying range of LLEs in accordance to Figure 2.

Based on the estimation of the LLEs, a Gaussian Bayesian classifier [26] is used to determine the probability of an orbit being regular, where we assume that LLEs are normally distributed in each class. First, the classifier is trained on LLEs resulting from 200 different initial conditions for T mapping iterations with the corresponding class labels resulting from the chosen reference being the generalized alignment index (GALI) [27]. Then, 10^4 test orbits are sampled from a regular grid in the range $[0, \pi] \times [0, 2\pi]$ with $\Delta q = \Delta p = 2\pi/10$, their LLE is calculated for T mapping iterations and the orbits are then classified. The results for $K = 0.9$ and $K = 2.0$ with $T = 50$ are shown in Figure 4, where the color map indicates the probability that the test orbit is regular. While for $K = 2.0$ the classifier provides a very clear distinction between regular and chaotic regions, the distinction between confined chaotic and regular orbits for $K = 0.9$ is less clear. With increasing number of mapping iterations, the number of misclassifications reduces as depicted in Figure 5. If the predicted probability that an orbit belongs to a certain class is lower than 70%, the prediction is not accepted and the orbit is marked as misclassified. With $K = 0.9$, the percentage of misclassified orbits does not drop below approximately 10%, because the transition between regular and chaotic motion is continuous.

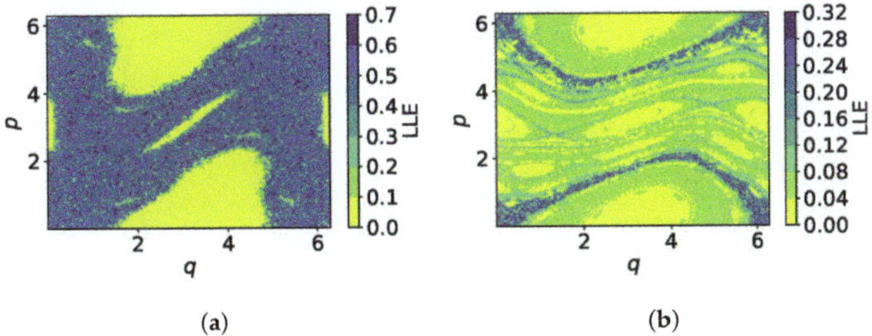

Figure 3. Local Lyapunov exponents in phase space of the standard map calculated with $T = 50$ mapping iterations for (**a**) $K = 2.0$, (**b**) $K = 0.9$.

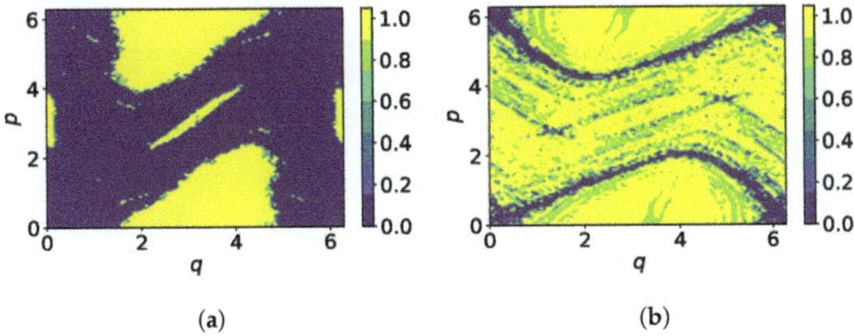

Figure 4. Orbit classification in standard map, (**a**) $K = 2.0$, (**b**) $K = 0.9$ for $T = 50$. The color map indicates the probability that the orbit is regular.

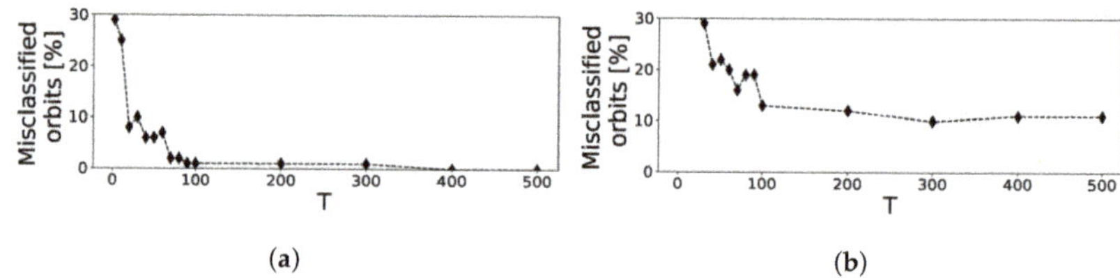

(a)   (b)

Figure 5. Percentage of misclassified orbits using a Bayesian classifier trained with 200 orbits for (**a**) $K = 2.0$ and (**b**) $K = 0.9$. 100 test orbits on an equally spaced grid in the range of $[0, \pi] \times [0, 2\pi]$ are classified as regular or chaotic depending on their LLE.

3.2. Sensitivity Analysis

The total Sobol' indices are calculated for the outputs from the symplectic surrogate model (Q, P) using Equation (15) with $N = 2000$ uniformly distributed random points within a box of size $[10^{-3} \times 10^{-3}]$ for each of the $T = 100$ mapping iterations as we are interested in the temporal evolution of the indices. For the standard map at $K = 0.9$ with $d = 2$ input and $D = 2$ output dimensions, 4 total Sobol' indices are obtained: S_q^Q and S_q^P denoting the influence of q and S_p^Q and S_p^P marking the influence of p on the output. We obtain good agreement with an MSE in the order of 10^{-6} between the indices obtained by the surrogate model and those using reference data.

As shown in Figure 6 for three different initial conditions for $K = 0.9$ depending on the orbit type, either chaotic or regular, the sensitivity indices behave differently. In case of a regular orbit close to a fixed point, S_j^i are oscillating, indicating that both input variables have similar influence on average. Getting further from the fixed point, closer to the border of stability, the influence of q gets bigger. This, however, is in contrast to the behavior in the chaotic case, where initially the variance in p has larger influence on the model output. However, when observing the indices over longer periods of time, both variables have similar influence. In Movie S01 in the supplemental material, the time evolution of all four total Sobol' indices obtained for the standard map are shown in phase space. Each frame is averaged over 10 subsequent mapping iterations. One snapshot is shown in Figure 7. The observation of the whole phase space sustains the findings in Figure 6.

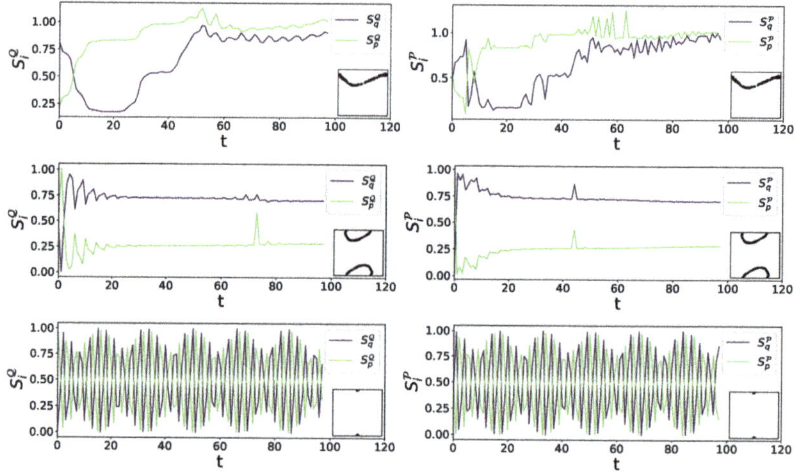

Figure 6. Total Sobol' indices as a function of time for three orbits of the standard map with $K = 0.9$—upper: chaotic orbit $(q, p) = (0.2, 5.6)$, middle: regular orbit $(q, p) = (1.76, 0.33)$, lower: regular orbit very close to fixed point $(q, p) = (\pi, 0.1)$.

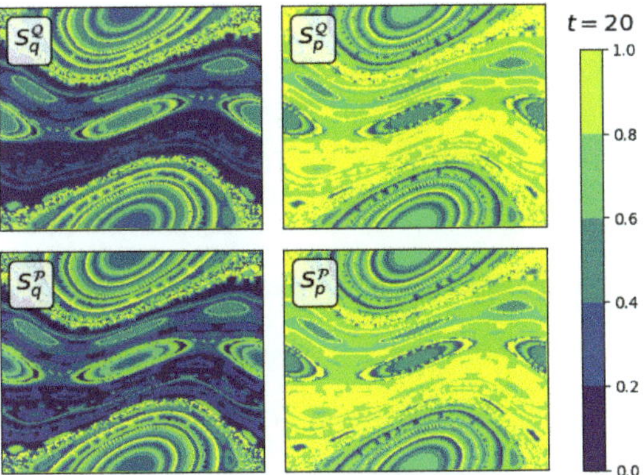

Figure 7. Total Sobol' indices (Equation (15)) for the standard map with $K = 0.9$ averaged from $t = 20$ to $t = 30$.

4. Conclusions

We presented an approach for orbit classification in Hamiltonian systems based on a structure preserving surrogate model combined with early classification based on local Lyapunov exponents directly available from the surrogate model. The approach was tested on two cases of the standard map. Depending on the perturbation strength, we either see a continuous transition from regular to chaotic orbits for $K = 0.9$ or a sharp separation between those two classes for higher perturbation strengths. This also impacts the classification results obtained from a Bayesian classifier. The presented method is applicable to chaotic Hamiltonian systems and is especially useful when a closed form expression for Poincaré maps is not available. Also, the accompanying sensitivity analysis provides valuable insight: in transition regions between regular and chaotic motion the Sobol' indices for time-series can be used to analyze the influence of input variables.

Author Contributions: Conceptualization, K.R., C.G.A., B.B. and U.v.T.; methodology, K.R., C.G.A., B.B. and U.v.T.; software, K.R.; validation, K.R., C.G.A., B.B. and U.v.T.; formal analysis, K.R., C.G.A., B.B. and U.v.T.; writing—original draft preparation, K.R.; visualization, K.R.; supervision, C.G.A., U.v.T. and B.B.; funding acquisition, C.G.A, B.B. and U.v.T. All authors have read and agreed to the published version of the manuscript.

Funding: The present contribution is supported by the Helmholtz Association of German Research Centers under the joint research school HIDSS-0006 "Munich School for Data Science-MUDS" and the Reduced Complexity grant No. ZT-I-0010.

Data Availability Statement: The data and source code that support the findings of this study are openly available [28] and maintained on https://github.com/redmod-team/SympGPR.

Conflicts of Interest: The authors declare no conflict of interest.

References

1. Ott, E. Chaos in Hamiltonian systems. In *Chaos in Dynamical Systems*, 2nd ed.; Cambridge University Press: Cambridge, UK, 2002; pp. 246–303. [CrossRef]
2. Lichtenberg, A.; Lieberman, M. *Regular and Chaotic Dynamics*; Springer: New York, NY, USA, 1992.
3. Albert, C.G.; Kasilov, S.V.; Kernbichler, W. Accelerated methods for direct computation of fusion alpha particle losses within, stellarator optimization. *J. Plasma Phys.* **2020**, *86*, 815860201. [CrossRef]
4. Eckmann, J.P.; Ruelle, D. Ergodic theory of chaos and strange attractors. *Rev. Mod. Phys.* **1985**, *57*, 617–656. [CrossRef]
5. Benettin, G.; Galgani, L.; Giorgilli, A.; Strelcyn, J. Lyapunov Characteristic Exponents for smooth dynamical systems and for Hamiltonian systems; A method for computing all of them. Part 1: Theory. *Meccanica* **1980**, *15*, 9–20. [CrossRef]
6. Benettin, G.; Galgani, L.; Giorgilli, A.; Strelcyn, J. Lyapunov Characteristic Exponents for smooth dynamical systems and for Hamiltonian systems; A method for computing all of them. Part 2: Numerical application. *Meccanica* **1980**, *15*, 21–30. [CrossRef]

7. Abarbanel, H.; Brown, R.; Kennel, M. Variation of Lyapunov exponents on a strange attractor. *J. Nonlinear Sci.* **1991**, *1*, 175–199. [CrossRef]
8. Abarbanel, H.D.I. Local Lyapunov Exponents Computed From Observed Data. *J. Nonlinear Sci.* **1992**, *2*, 343–365. [CrossRef]
9. Eckhardt, B.; Yao, D. Local Lyapunov exponents in chaotic systems. *Phys. D Nonlinear Phenom.* **1993**, *65*, 100–108. [CrossRef]
10. Amitrano, C.; Berry, R.S. Probability distributions of local Liapunov exponents for small clusters. *Phys. Rev. Lett.* **1992**, *68*, 729–732. [CrossRef] [PubMed]
11. Arnold, V. *Mathematical Methods of Classical Mechanics*; Springer: New York, NY, USA, 1989; Volume 60.
12. Rath, K.; Albert, C.G.; Bischl, B.; von Toussaint, U. Symplectic Gaussian process regression of maps in Hamiltonian systems. *Chaos* **2021**, *31*, 053121. [CrossRef]
13. Skokos, C. The Lyapunov Characteristic Exponents and Their Computation. In *Dynamics of Small Solar System Bodies and Exoplanets*; Springer: Berlin/Heidelberg, Germany, 2010; pp. 63–135. [CrossRef]
14. Sobol, I. Global sensitivity indices for nonlinear mathematical models and their Monte Carlo estimates. *Math. Comput. Simul.* **2001**, *55*, 271–280. [CrossRef]
15. Sobol, I. On sensitivity estimation for nonlinear math. models. *Matem. Mod.* **1990**, *2*, 112–118.
16. Saltelli, A.; Annoni, P.; Azzini, I.; Campolongo, F.; Ratto, M.; Tarantola, S. Variance based sensitivity analysis of model output. Design and estimator for the total sensitivity index. *Comput. Phys. Commun.* **2010**, *181*, 259–270. [CrossRef]
17. Chirikov, B.V. A universal instability of many-dimensional oscillator systems. *Phys. Rep.* **1979**, *52*, 263–379. [CrossRef]
18. Rasmussen, C.E.; Williams, C.K.I. *Gaussian Processes for Machine Learning*; MIT Press: Cambridge, MA, USA, 2005.
19. Álvarez, M.A.; Rosasco, L.; Lawrence, N.D. Kernels for Vector-Valued Functions: A Review. *Found. Trends Mach. Learn.* **2012**, *4*, 195–266. [CrossRef]
20. Solak, E.; Murray-smith, R.; Leithead, W.E.; Leith, D.J.; Rasmussen, C.E. Derivative Observations in Gaussian Process Models of Dynamic Systems. In *NIPS Proceedings 15*; Becker, S., Thrun, S., Obermayer, K., Eds.; MIT Press: Cambridge, MA, USA, 2003; pp. 1057–1064.
21. Eriksson, D.; Dong, K.; Lee, E.; Bindel, D.; Wilson, A. Scaling Gaussian process regression with derivatives. In *Advances in Neural Information Processing Systems 31: Annual Conference on Neural Information Processing Systems 2018 (NeurIPS 2018), Montréal, QC, Canada, 3–8 December 2018*; Curran Associates, Inc.: Red Hook, NY, USA, 2018; pp. 6868–6878.
22. Sudret, B. Global sensitivity analysis using polynomial chaos expansions. *Reliab. Eng. Syst. Saf.* **2008**, *93*, 964–979. [CrossRef]
23. Marrel, A.; Iooss, B.; Laurent, B.; Roustant, O. Calculations of Sobol indices for the Gaussian process metamodel. *Reliab. Eng. Syst. Saf.* **2009**, *94*, 742–751. [CrossRef]
24. Geist, K.; Parlitz, U.; Lauterborn, W. Comparison of Different Methods for Computing Lyapunov Exponents. *Prog. Theor. Phys.* **1990**, *83*, 875–893. [CrossRef]
25. Ellner, S.; Gallant, A.; McCaffrey, D.; Nychka, D. Convergence rates and data requirements for Jacobian-based estimates of Lyapunov exponents from data. *Phys. Lett. A* **1991**, *153*, 357–363. [CrossRef]
26. Pedregosa, F.; Varoquaux, G.; Gramfort, A.; Michel, V.; Thirion, B.; Grisel, O.; Blondel, M.; Prettenhofer, P.; Weiss, R.; Dubourg, V.; et al. Scikit-learn: Machine Learning in Python. *J. Mach. Learn. Res.* **2011**, *12*, 2825–2830.
27. Skokos, C.; Bountis, T.; Antonopoulos, C. Geometrical properties of local dynamics in Hamiltonian systems: The Generalized Alignment Index (GALI) method. *Phys. D Nonlinear Phenom.* **2007**, *231*, 30–54. [CrossRef]
28. Rath, K.; Albert, C.; Bischl, B.; von Toussaint, U. SympGPR v1.1: Symplectic Gaussian process regression. *Zenodo* **2021**. [CrossRef]

physical sciences forum

Proceeding Paper

The ABC of Physics [†]

John Skilling [1,‡] and Kevin Knuth [2,*,‡]

1. Maximum Entropy Data Consultants, Cambridge CB4 1XE, UK; john@skilling.co.uk
2. Department of Physics, University at Albany (SUNY), 1400 Washington Ave, Albany, NY 12222, USA
* Correspondence: kknuth@albany.edu
† Presented at the 40th International Workshop on Bayesian Inference and Maximum Entropy Methods in Science and Engineering, online, 4–9 July 2021.
‡ These authors contributed equally to this work.

Abstract: Why quantum? Why spacetime? We find that the key idea underlying both is *uncertainty*. In a world lacking probes of unlimited delicacy, our knowledge of quantities is necessarily accompanied by uncertainty. Consequently, physics requires a calculus of number *pairs* and not only scalars for quantity alone. Basic symmetries of shuffling and sequencing dictate that pairs obey ordinary component-wise addition, but they can have three different multiplication rules. We call those rules A, B and C. "A" shows that pairs behave as complex numbers, which is *why quantum theory is complex*. However, consistency with the ordinary scalar rules of probability shows that the fundamental object is not a particle on its Hilbert sphere but a stream represented by a Gaussian distribution. "B" is then applied to pairs of complex numbers (qubits) and produces the Pauli matrices for which its operation defines the space of four vectors. "C" then allows integration of what can then be recognised as energy-momentum into time and space. The picture is entirely consistent. *Spacetime is a construct of quantum* and not a container for it.

Keywords: quantum; spacetime; uncertainty; symmetry

Citation: Skilling, J.; Knuth, K.H. The ABC of Physics. *Phys. Sci. Forum* **2021**, *3*, 9. https://doi.org/10.3390/psf2021003009

Academic Editors: Wolfgang von der Linden and Sascha Ranftl

Published: 10 December 2021

1. Strategy

Simplicity yields generality, and generality is power. There are deep mysteries in physics, such as why space has three dimensions and why quantum formalism is complex. Inquiry at such depth demands sparse assumptions of compelling generality. We aim to leave no room for plausible doubt; thus, we allow no delicate assumptions (such as continuity, perhaps) which could plausibly be denied. We paint with a broad brush, aiming to mirror simplicity of ideas with simplicity of presentation. Our aim is to expose the inevitable *language* of physics in a form accessible to neophyte students as well as experienced professionals. This is only a beginning, and we do not proceed to discuss the physical laws that form content of the language.

In attempting to understand the world, we seem forced to think of it in terms of adequately isolated parts with adequately separable properties. To obtain generality, we hypothecate symmetries such that laws applying to one part will apply to another. Otherwise, without consistent rules, we will find no generalities, and our quest will fail. Symmetries are our only hope. If the world is to be comprehended at all, this is the path we should follow. What, if anything, the symmetries apply to will only be apparent later when we try to match our intellectual constructions to sensory impressions of the world.

2. Mathematics

We first suppose a "with" operator that links two parts ("stuff") to make compound stuff. Technically, this operator is deemed to possess *closure*: stuff-with-stuff = stuff. Furthermore, since we have closure, we can continue adding other stuff to produce yet more stuff. For this to be fully useful, we suppose that we can shuffle stuff around without it making any difference. Formally, shuffling is associative commutativity.

$$A\text{-with-}(B\text{-with-}C) = (A\text{-with-}B)\text{-with-}C \quad \text{associative}$$
$$A\text{-with-}B = B\text{-with-}A \quad \text{commutativity} \tag{1}$$

If our stuff has only one property of interest, it is not hard to show that the "with" operator can without loss of generality be taken to be standard arithmetical addition [1–4]. Any other representation is isomorphic. In order to model an arbitrarily large world, we suppose that combination can be conducted indefinitely without repeat: $X\text{-with-}Y \neq X$ unless Y is null. Otherwise, we would obtain the wrap around for integers limited to finite storage. This is why stuff adds up [5].

$$a + (b + c) = (a + b) + c \quad \text{associative}$$
$$a + b = b + a \quad \text{commutativity} \quad \Big\} \quad (\textit{sum rule}) \tag{2}$$

It is not difficult: Children understand it as they learn to count. The wide utility of addition mirrors the ubiquity of symmetry under shuffling. There are many applications, but we might think of shuffling as a parallel operation which could take place in some sort of proto-space.

Now, we will want "+" to continue to work regardless of who supplies the stuff or, for that matter, who receives it; thus, we next suppose a "then" operator that transfers stuff, possibly with modification, as in $A\text{-then-}B$. This operator is also supposed to possess closure: $A\text{-then-}B\text{-then-}$ is itself a transfer $C\text{-then-}$ so that suppliers can be chained, and as before we suppose that chaining can be performed indefinitely. We might think of -then- as a series operation that could take place in some sort of proto-time.

For -with- to be additive regardless of suppliers or receivers, we need -then- to be distributive.

$$A\text{-then-}(B\text{-with-}C) = (A\text{-then-}B)\text{-with-}(A\text{-then-}C) \quad \text{left distributive}$$
$$(A\text{-with-}B)\text{-then-}C = (A\text{-then-}C)\text{-with-}(B\text{-then-}C) \quad \text{right distributive} \tag{3}$$

With only one property of interest in play, the representation is merely scalar. The only freedom allowed within the sum-rule convention is scaling; thus, the representation "·" of -then- has to be standard arithmetical multiplication, possibly scaled by some constant γ [3,4]:

$$a \cdot b = \gamma ab \quad (\textit{product rule}) \tag{4}$$

which is most commonly set by convention to one. Again, this is not difficult: children learn it informally as they are taught multiplication, and they learn γ with percentages.

Note that we are building arithmetic and not Boolean calculus. Our "then" describes sequencing and is distributive over "with" but not the other way around. We point out that only with standard arithmetic in place can we build the sophisticated mathematics, which gives us quantified science. This is why mathematics is so effective—we have *designed* it to obey the fundamental symmetries which are required for our generalisations. Moreover, we need the symmetries, for denial would remove our mathematics and render us powerless. How could we justify Hilbert spaces if we cannot even add up?

3. Physics

Physics is more subtle, for arbitrary precision is not available with finite probes (Figure 1, left).

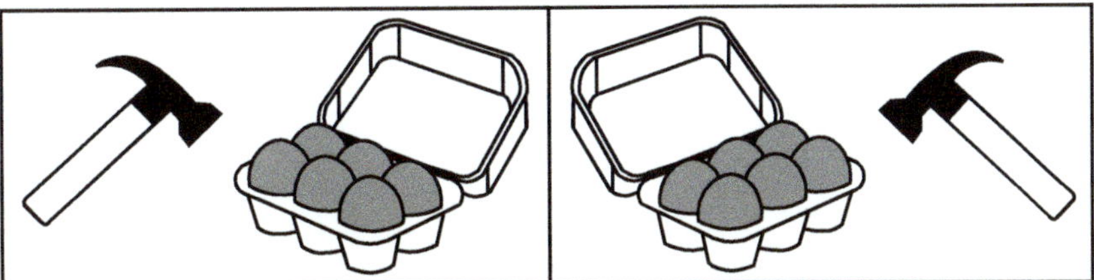

Figure 1. (**Left**) Uncertainty: probe and fragile target. (**Right**) Measurement: delicate probe with target.

This means that any particular quantity is necessarily associated with an uncertainty. Hence, a minimal description (and anything more would lack rationale and be mathematically redundant) requires a *pair* of numbers to quantify a property. We do not mean "$\mu \pm \sigma$," which is merely crude shorthand for a distribution $\Pr(x)$ over the available values x—in which those numbers too would be uncertain, resulting in indefinite regress. The fundamentally irreducible pair-wise connection is simpler and more intimate than that. Uncertainty becomes part of the very *language* of physics and is impossible to remove.

To define the connections analogous to scalar addition and multiplication, we start by imposing the same ubiquitous symmetries as before. Associative commutativity yields component-wise addition for -with-:

$$\begin{pmatrix} a_1 \\ a_2 \end{pmatrix} + \begin{pmatrix} b_1 \\ b_2 \end{pmatrix} = \begin{pmatrix} a_1 + b_1 \\ a_2 + b_2 \end{pmatrix} \tag{5}$$

and distributivity yields bilinear multiplication for -then-:

$$\begin{pmatrix} a_1 \\ a_2 \end{pmatrix} \cdot \begin{pmatrix} b_1 \\ b_2 \end{pmatrix} = \begin{pmatrix} \gamma_1 a_1 b_1 + \gamma_2 a_1 b_2 + \gamma_3 a_2 b_1 + \gamma_4 a_2 b_2 \\ \gamma_5 a_1 b_1 + \gamma_6 a_1 b_2 + \gamma_7 a_2 b_1 + \gamma_8 a_2 b_2 \end{pmatrix} \tag{6}$$

but now with eight apparently arbitrary coefficients γ and not only a single removable scale factor. It is true that we have four degrees of freedom to choose coordinates, but that still leaves four of the γ's free so that multiplication is not yet adequately defined. This extra subtlety of physics requires us to make the distributivity *associative*, meaning that the effect of chaining does not depend on how the factors are grouped into compounds.

$$a \cdot (b \cdot c) = (a \cdot b) \cdot c \qquad \text{associative} \tag{7}$$

In one dimension, associativity was automatically an emergent property of multiplication, but we need to impose it in two dimensions. Associative multiplication of pairs yields quadratic equations for the γ's for which its multiple solutions A,B,C,D,E and F enable the rich framework of physics. In summary, the following is the case.

$$\begin{array}{lcl} \text{Quantifiable physics} & \Longrightarrow & \text{quantity-\&-uncertainty } \textit{pairs} \\ \text{Associative commutativity (shuffling)} & \Longrightarrow & \text{component-wise sum rule } (+) \\ \text{Associative distributivity (sequencing)} & \Longrightarrow & \text{product rules A,B,C,D,E,F } (\cdot) \end{array} \tag{8}$$

Explicitly, the six product rules are the following [6–8].

$$\begin{pmatrix} a_1 \\ a_2 \end{pmatrix} \cdot \begin{pmatrix} b_1 \\ b_2 \end{pmatrix} =$$

$$\underbrace{\begin{pmatrix} a_1 b_1 - a_2 b_2 \\ a_1 b_2 + a_2 b_1 \end{pmatrix}}_{A} \text{ or } \underbrace{\begin{pmatrix} a_1 b_1 + a_2 b_2 \\ a_1 b_2 + a_2 b_1 \end{pmatrix}}_{B} \text{ or } \underbrace{\begin{pmatrix} a_1 b_1 \\ a_1 b_2 + a_2 b_1 \end{pmatrix}}_{C} \text{ or } \underbrace{a_1 \begin{pmatrix} b_1 \\ b_2 \end{pmatrix}}_{D} \text{ or } \underbrace{\begin{pmatrix} a_1 \\ a_2 \end{pmatrix} b_1}_{E} \text{ or } \underbrace{\begin{pmatrix} a_1 b_1 \\ 0 \end{pmatrix}}_{F} \tag{9}$$

The first three, A,B and C, are pair-pair products, the next two, D and E, are degenerate scalar-pair products, and the last F is ordinary scalar–scalar multiplication appearing as the doubly degenerate case.

4. Product Rule F

It was inevitable that two-dimensional multiplication would include one-dimensional scalar multiplication as a special case, and what our derivation demonstrates is that there is necessary consistency between one and two dimensions. Both follow from the same basic symmetries of shuffling and sequencing. In one dimension, the scalar sum and product rules are those of probability, observed from this viewpoint as part of an intellectual structure common to both mathematics and physics.

Whether quantum or classical, physics makes predictions expressed as *likelihoods* $\Pr(\text{outcome} \mid \text{setup})$, assuming what we think we know about the setup, expressed as the *prior*. Bayesian inference then uses actual outcomes to refine the predictions (the *posterior*) and assess the predictive quality of our assumptions (the *evidence*). Probability and quantum theory (which is basic physics) share a common foundation, and quantum behaviour fits seamlessly into Bayesian analysis no differently to anything else. There can be no quantum weirdness in this approach. It is all only ordinary inference.

Note our bottom-up strategy. Our assumptions refer to basic symmetries of only two or three items at a time. We start with $1 + 1 = 2$ items and build up from there. This is opposite to approaches of superficially greater sophistication which specify top-down from infinity $\sum_1^\infty (\cdot)$. That strategy is deliberate. Infinity and the continuum involve delicate limits that might not hold in all circumstances. We think it perversely fragile to traverse delicate analysis in order to return, ultimately, to discrete predictions which never needed the delicacy in the first place.

5. Product Rule A

Product rule A is complex multiplication, which we can write in operator form as follows.

$$a \cdot = \begin{pmatrix} a_1 & -a_2 \\ a_2 & a_1 \end{pmatrix} = re^{i\theta} \tag{10}$$

The condition $r = 1$ selects operators which under repeated application $(a \cdot)^n$ cause neither divergence towards infinity $(r > 1)$ nor collapse towards zero $(r < 1)$. Thus, $r = 1$ defines *unit quantity*. Knowledge of quantity is invariant under such unit-modulus operations, which add constants to the phase of the operand while leaving its modulus unchanged. Correspondingly, our knowledge $\Pr(\phi)$ of the phase of an operand is invariant to offset, obeying $\Pr(\phi) = \Pr(\phi + \theta)$ for any θ, indicating uniformity.

$$\Pr(\phi) = \frac{1}{2\pi} = \text{uniform over } \phi \in [0, 2\pi) \tag{11}$$

Later, it will be observed that neither of the other pair–pair product rules can support a proper prior distribution, and *this is why quantum theory uses complex numbers*. It is the only way of enabling consistency with probability. Uncertainty being identified with phase, the modulus becomes associated with quantity, and the unit modulus represents unit quantity.

Incidentally, phase has to be a continuous variable. If it was not, any gap could be filled in by adding complex numbers from either side: one from the upper boundary and one from the lower in order to obtain a sum with intermediate phase. Whilst absolute phase of any newly introduced quantity is unknown (and uniformly distributed), *relative* phase is the critical ingredient that distinguishes representation of quantity-with-uncertainty from representation of quantity alone. Relative phase manifests as interference.

6. Born Rule

When we combine objects, the scalar sum rule demands that we add quantity. That is the definition: quantity is what adds up. So what scalar property of complex numbers could be additive? Complex moduli do not add up, either directly $|a + b| \neq |a| + |b|$ or through any function $f(|a + b|) \neq f(|a|) + f(|b|)$. Phase differences interfere and appear to force an inconsistency with scalar addition. However, we do not know the phases of independent objects; thus, we do not actually know what the interference should be. Fortunately, the rules of probability instruct us to average ("marginalise") over what we do not know in order to arrive at the reproducible behaviour that we seek. Moreover, we *can* attain consistency on average, $\langle f(|a + b|) \rangle = \langle f(|a|) \rangle + \langle f(|b|) \rangle$, provided we use $f(x) = x^2$. That modulus-squared form of the following:

$$\left\langle \, |\, ae^{i\alpha} + be^{i\beta} \, |^2 \right\rangle_{\alpha,\beta} = \left\langle \, |\, ae^{i\alpha} \, |^2 \right\rangle_{\alpha} + \left\langle \, |\, be^{i\beta} \, |^2 \right\rangle_{\beta} = |a|^2 + |b|^2 \tag{12}$$

is forced upon us and yields a general additivity that holds for any number of components. The average-modulus-squared relationship is the archetype of the *Born rule* of quantum theory [9]. By updating our nomenclature of complex numbers to the more traditional ψ, we have additive scalar quantification $\langle |\psi|^2 \rangle$, which we interpret as follows.

$$Q = \langle |\psi|^2 \rangle = \text{rate of supply of } \psi = \text{intensity.} \tag{13}$$

This is how uncertainty manifests in the formalism. The observably additive intensity emitted by sources and observed by receivers is a modulus-squared *average*. Averaging can be performed artificially by picturing an ensemble of possibilities, as expressed either algebraically with a formula for $\Pr(\cdot)$ or arithmetically as Monte Carlo samples. In the laboratory, averaging is performed by repetition. Thus, *the fundamental object of inquiry is a stream* and not, as commonly assumed, a particle. A particle is less fundamental because it carries the extra information that a detector had fired. Streams add up. Particles only add up if they are independent because otherwise phase difference is liable to cause interference so that 1-with-1 could become anything from 0 to 4.

Observing Q yields information about ψ partially because the phase remains unknown, and the observation is only an average. The probability distribution for ψ under this quadratic constraint Q is assigned by maximum entropy as the Gaussian:

$$\Pr(\psi \mid Q) = \frac{\exp(-|\psi|^2/Q)}{\pi Q} \tag{14}$$

just as standard in classical inference and signal processing.

The foundations of quantum theory begin to become apparent, but so far all we have quantified is existence. We need more. We need properties.

7. Product Rule B

Product rule B is hyperbolic (or "split-complex") multiplication, written in operator form as follows.

$$a \cdot = \begin{pmatrix} a_1 & a_2 \\ a_2 & a_1 \end{pmatrix} \quad \text{with} \quad r = \sqrt{|\det(a \cdot)|} \tag{15}$$

The condition $r = 1$ selects operators which under repeated application $(a \cdot)^n$ cause neither divergence towards infinity ($r > 1$) nor collapse towards zero ($r < 1$). Thus, $r = 1$ defines *unit quantity*. Instead of complex phase $\arctan(a_2/a_1)$, we have pseudo-phase $\phi = \operatorname{arctanh}(a_2/a_1)$. Offsets of ϕ leave r unchanged, but the range is unbounded $\phi \in (-\infty, \infty)$. Hence, as anticipated, we could not use ϕ to express uncertainty because the corresponding uniform prior would be improper. Rule B does not admit some alternative representation of quantity and uncertainty.

Instead, we upgrade our inquiry to binary this-or-that *properties*, each of which can exist or not and is represented by a quantity and uncertainty pair. We call the binaries of the following:

$$\psi = \begin{pmatrix} \psi_\uparrow \\ \psi_\downarrow \end{pmatrix} = \begin{pmatrix} \psi_0 + i\psi_1 \\ \psi_2 + i\psi_3 \end{pmatrix} \tag{16}$$

qubits. The detection of $\langle |\psi_\uparrow|^2 \rangle$ by a $\uparrow$-detector would signify the supply of $\uparrow$ and similarly for $\downarrow$ so that $\langle |\psi_\uparrow|^2 + |\psi_\downarrow|^2 \rangle$ would signify $\uparrow$-or-$\downarrow$ presence of qubits with either property.

Shuffling and sequencing are still to be obeyed, and the ABCDEF rules still apply in the complex field. So far, we are invoking rules A and B. For the sake of simplicity, we restrict attention to unit operators and write them in generator form. The generator form of the following:

$$a \cdot = \exp(\phi G) = \lim_{n \to \infty} \left(1 + \frac{G}{n} \right)^{n\phi} \tag{17}$$

allows us to separate the (complex) magnitude ϕ of a multiplication $(a \cdot)$ from its structure G. From (9), the unit operators for A and B are as follows:

$$\begin{pmatrix} \cos\phi & -\sin\phi \\ \sin\phi & \cos\phi \end{pmatrix} = \exp(\phi \mathbf{A}) \quad \text{where} \quad \mathbf{A} = \begin{pmatrix} 0 & -1 \\ 1 & 0 \end{pmatrix}$$
$$\begin{pmatrix} \cosh\phi & \sinh\phi \\ \sinh\phi & \cosh\phi \end{pmatrix} = \exp(\phi \mathbf{B}) \quad \text{where} \quad \mathbf{B} = \begin{pmatrix} 0 & 1 \\ 1 & 0 \end{pmatrix} \tag{18}$$

where we recognise the Hermitian *Pauli matrices*:

$$\sigma_0 = \underbrace{\begin{pmatrix} 1 & 0 \\ 0 & 1 \end{pmatrix}}_{1}, \quad \sigma_x = \underbrace{\begin{pmatrix} 0 & 1 \\ 1 & 0 \end{pmatrix}}_{\mathbf{B}}, \quad -\sigma_y = \underbrace{\begin{pmatrix} 0 & -i \\ i & 0 \end{pmatrix}}_{i\mathbf{A}}, \quad \sigma_z = \underbrace{\begin{pmatrix} 1 & 0 \\ 0 & -1 \end{pmatrix}}_{\mathbf{BA}} \tag{19}$$

that mix trigonometric and hyperbolic rotations in complex context.

7.1. Quantification

For quantification, we still have the Born rule, initially as individual averages of $|\psi_\uparrow|^2$ and $|\psi_\downarrow|^2$, but quickly promoted to $\langle \psi^\dagger \mathbf{H}\, \psi \rangle$ for any Hermitian $\mathbf{H}$. Specifically, the Pauli matrices yield scalar observables.

$$\begin{aligned}
p_0 &= \langle \psi^\dagger \sigma_0\, \psi \rangle = \langle \psi_0^2 + \psi_1^2 + \psi_2^2 + \psi_3^2 \rangle \\
p_x &= \langle \psi^\dagger \sigma_x\, \psi \rangle = 2\langle \psi_0\psi_2 + \psi_1\psi_3 \rangle \\
p_y &= \langle \psi^\dagger \sigma_y\, \psi \rangle = 2\langle \psi_0\psi_3 - \psi_1\psi_2 \rangle \\
p_z &= \langle \psi^\dagger \sigma_z\, \psi \rangle = \langle \psi_0^2 + \psi_1^2 - \psi_2^2 - \psi_3^2 \rangle
\end{aligned} \tag{20}$$

Of these, p_0 observes qubits as a whole. Given a covariance matrix $\rho = \langle \psi\, \psi^\dagger \rangle$ derived from such observations (which in practice would often include statistical uncertainty too), the maximum entropy assignment of probability in any dimension is the Gaussian.

$$\Pr(\psi \mid \rho) = \frac{\exp(-\psi^\dagger \rho^{-1} \psi)}{\det(\pi \rho)} \tag{21}$$

This distribution of ψ can be used to predict observable quantities in the usual manner that is no differently from any other application of probability. Knowing what a system is enables us to predict how it will behave, whether that is probabilistic or definitive.

It is clear from (21) that initial coordinates could be replaced by any unitary combination.

$$\Pr(\psi \mid \rho) = \frac{\exp(-(\mathbf{U}\psi)^\dagger (\mathbf{U}\rho\mathbf{U}^\dagger)^{-1}(\mathbf{U}\psi))}{\det(\pi \rho)}, \qquad (\mathbf{U}^\dagger \mathbf{U} = 1) \tag{22}$$

Hence, there is no observational test of whether ψ or $\mathbf{U}\psi$ is the fundamental representation; thus, our physics needs to be invariant under unitary transformation. This gives the complex representations of quantum theory flexibility beyond that accessible to classical scalars for which quantity is restricted to positive values.

7.2. Quantisation

We become aware of quantisation when a probe initialised in a fragile metastable state is brought to the target stream (Figure 1(right)). If the target triggers effectively irreversible descent of the probe into a disordered state (cleverly engineered to be a macroscopic pointer angle or suchlike), then that becomes available as an essentially permanent macroscopic bit of information signifying that something (a quantum) had been present in the target stream at that time. It is from this point that the experimenter may in person or by proxy become aware of the digital event and incorporate it into probabilistic modelling.

Assuming that the interaction had been engineered to preserve the identities of target and probe, then the something (the quantum) could then be intercepted from the ongoing stream and preserved for later use. Its total modulus would then be known (and conventionally assigned as 1) so that its "wave function" ψ would be represented by the distribution:

$$\Pr(\psi) \propto \delta(\psi^\dagger \psi - 1) \tag{23}$$

with known modulus and unknown phase. With the wave function ψ thereby normalised and confined to the Hilbert sphere $\psi^\dagger \psi = 1$, the covariance ρ is known as the *density matrix*. However, it may be noted that such confinement damages the smooth elegance of the Gaussian analysis of streams.

This is the source of entanglement, contextuality and such quintessentially quantum phenomena. It is simply Bayesian analysis in the unfamiliar context of complex numbers (as suggested by [10]) and in indirect modulus-squared observation, usually written in the Dirac bracket notation with $\psi = |\psi\rangle$ and $\psi^\dagger = \langle\psi|$, that was developed for physics [11,12] before the Bayesian paradigm became pre-eminent in inference.

For example, if p_z were observed equal to p_0, then the definitions (20) would force ψ to be purely $\psi_\uparrow$, an eigenvector of σ_z (with unknown phase) with covariance $\rho = \psi_\uparrow \psi_\uparrow^\dagger$. Subsequent prediction of ψ with (21) would ensure that p_z remained equal to p_0. Measurements of p_x, on the other hand, would average to zero upon repetition and restriction to the Hilbert sphere (23) of an individual quantum would force individual outcomes to take either one of the eigenvectors of σ_x at random, incidentally destroying the memory of the earlier p_z. We see that "wave-function collapse" is just the incorporation of new data into ordinary Bayesian inference on the Hilbert sphere.

7.3. Transformations

The Pauli matrices generate the 6-parameter *Lorentz group* of unit operators.

$$\exp(\phi_x \sigma_x + \phi_y \sigma_y + \phi_z \sigma_z) \quad \text{with complex coefficients } \phi = -\tfrac{1}{2}(\xi + i\eta) \tag{24}$$

On using η_z only, the Pauli observables transform as follows:

$$\begin{pmatrix} p_0' \\ p_x' \\ p_y' \\ p_z' \end{pmatrix} = \begin{pmatrix} 1 & 0 & 0 & 0 \\ 0 & \cos\eta & -\sin\eta & 0 \\ 0 & \sin\eta & \cos\eta & 0 \\ 0 & 0 & 0 & 1 \end{pmatrix} \begin{pmatrix} p_0 \\ p_x \\ p_y \\ p_z \end{pmatrix} \tag{25}$$

while on using ξ_z only, they transform as follows.

$$\begin{pmatrix} p_0' \\ p_x' \\ p_y' \\ p_z' \end{pmatrix} = \begin{pmatrix} \cosh\xi & 0 & 0 & -\sinh\xi \\ 0 & 1 & 0 & 0 \\ 0 & 0 & 1 & 0 \\ -\sinh\xi & 0 & 0 & \cosh\xi \end{pmatrix} \begin{pmatrix} p_0 \\ p_x \\ p_y \\ p_z \end{pmatrix} \tag{26}$$

In each case, we have the following:

$$p_0^2 - p_x^2 - p_y^2 - p_z^2 = m^2 > 0 \text{ is invariant} \tag{27}$$

and the axis could have been in any (x, y, z) direction. We recognise the rotation and boost behaviour of 4-momentum, but the identification would be premature because we do not yet have spacetime.

8. Product Rule C

Product rule C in operator form is as follows.

$$a \cdot = \begin{pmatrix} a_1 & 0 \\ a_2 & a_1 \end{pmatrix} \tag{28}$$

As before, $r = \sqrt{|\det(a \cdot)|} = 1$ defines *unit quantity*. Instead of complex phase $\arctan(a_2/a_1)$, we have pseudo-phase $t = a_2/a_1$. Offsets of t leave r unchanged, but the range is unbounded $t \in (-\infty, \infty)$. Hence, as anticipated, we could not use t to express uncertainty because the uniform prior would be improper. The only choice really was rule A.

Unit operations now take the following form.

$$\begin{pmatrix} 1 & 0 \\ t & 1 \end{pmatrix} = \exp(t\mathbf{C}) \quad \text{where} \quad \mathbf{C} = \begin{pmatrix} 0 & 0 \\ 1 & 0 \end{pmatrix} \tag{29}$$

This acts as an integrator:

$$\begin{pmatrix} 1 & 0 \\ t & 1 \end{pmatrix} \begin{pmatrix} u \\ U \end{pmatrix} = \begin{pmatrix} u \\ U + ut \end{pmatrix} \tag{30}$$

so that U is the integral of u with $dU = u\, dt$. Note that $\mathbf{C}$ operates in the real domain. It is not Hermitian, so it applies to scalar operands and not to complex entities such as ψ, and its coefficient t will be real. Rule C meaningfully applies to (real-valued) phase, rephasing ψ to $\psi e^{-i\theta}$ (the minus sign is conventional).

Take a qubit ψ with invariant m. Define $d\tau$ through $\boxed{d\theta = m\, d\tau}$ by using m as a clock. Under Pauli, we have observed that m transforms as the 4-vector (p_0, p_x, p_y, p_z), which is covariant by convention. Phase θ is a 2π-periodic scalar that cannot be rescaled because its 2π-periodicity represents *interference*—the distinguishing observational feature of quantum theory. Hence, $d\tau$ must transform as a contravariant 4-vector (dt, dx, dy, dz) with invariant of the following:

$$(d\tau)^2 = (dt)^2 - (dx)^2 - (dy)^2 - (dz)^2 \tag{31}$$

obeying the following:

$$d\theta = (p_0 dt - p_x dx - p_y dy - p_z dz)/\hbar \tag{32}$$

where the arbitrary constant $\hbar$ (which is best set to one) defines our units of t and $\mathbf{x}$ relative to m and $\mathbf{p}$. With ψ being rephased by $e^{-i\theta}$, the Schrödinger equations of the following:

$$i\hbar\frac{\partial\psi}{\partial t} = p_0\psi, \quad i\hbar\frac{\partial\psi}{\partial x} = -p_x\psi, \quad i\hbar\frac{\partial\psi}{\partial y} = -p_y\psi, \quad i\hbar\frac{\partial\psi}{\partial z} = -p_z\psi. \tag{33}$$

are immediate.

All that remains is to recognise the symbols of the following.

θ	action		
m	rest mass	τ	proper time
p_0	energy	t	time
(p_x, p_y, p_z)	momentum $\mathbf{p}$	(x, y, z)	location $\mathbf{x}$

Moreover, we must identify the Lorentz coefficients $\phi = -\frac{1}{2}(\xi + i\eta)$ within Minkowski spacetime.

v	velocity dx/dt
$\cosh \xi$	Lorentz factor $(1 - v^2)^{-1/2}$
η	rotation angle

8.1. Viewpoint

Traditionally, energy and momentum are obtained from the Schrödinger equations as differentials with respect to time and space. With ABC, it is the other way round—time and space emerge from quantum theory as integrals of energy and momentum.

$$(p_0, p_x, p_y, p_z) \underset{\text{traditional}}{\overset{\text{ABC}}{\rightleftharpoons}} (t, x, y, z)$$

Spacetime is a construct of quantum, not a container for it.

The ABC derivation dictates why space has three dimensions—and that is the right way round. Space and time are measured by theodolites and clocks, and observations are at root quantum in nature.

We have now developed a unified foundation based on the necessary incorporation of uncertainty with quantity.

Associative commutativity Associative distributivity	$\overset{\text{A}}{\Longrightarrow}$	Probability
	$\overset{\text{AB}}{\Longrightarrow}$	Probability & Quantum
	$\overset{\text{ABC}}{\Longrightarrow}$	Probability & Quantum & Spacetime

Further development would include formalising qubit–qubit interactions (two Pauli matrices) by using the Dirac equation and continuing to quantum field theory. That would allow theoretical demonstration of idealised measurement probes and of how observation can be used both to predict future evolution and to retrodict past behaviour. However, since ABC recovers the standard equations, new insight would not be immediately expected.

9. Speculations

With a consistent formulation based on so few and general assumptions, it seems almost inconceivable that inconsistency could arise in future developments. Would a user wish to deny uncertainty, or shuffling, or sequencing?

However, there is more to be conducted. Although spacetime has been constructed with a locally Minkowski metric, inheriting continuity from complex phase, ABC does not require the manifold to be flat. We should presumably expect that curvature will follow physical matter as general relativity indicates. Yet gravitational singularities appear to be inconsistent with the required reversibility of quantum formalism.

One may ask for the source of that reversibility, for it is at least superficially plausible that dropping an object into a black hole is *not* a reversible process—at least not on a timescale of interest to the experimenter. Since it appears that the fundamental object is not a particle on its Hilbert sphere but a stream represented by a Gaussian distribution, it seems that there is scope for revisiting this question from a revised viewpoint.

Traditional developments of quantum theory and of geometry have accepted the notion of a division algebra [13,14] in which every operation is presumed to have an inverse. However, that may not be true. ABC does not need to assume inverses. They might fail in extreme situations. Assuming division algebras may have been an assumption too far.

One may also speculate on the roles of the unused product rules D and E. Traditionally, these scalar-vector product rules are imposed as part of the axiomatic structure of a vector space. Users are just expected to accept them on the basis of initial plausibility, familiarity and submission to conventional authority. However, ABC did not need to assume them. In the two-parameter quantity and uncertainty space, D and E appear as additional candidate rules. Each of A, B, C, and F had critically important roles. Why would D and E not have critically important roles too?

Inquiry at this depth does not throw up results that we should casually discard. Rules D and E suggest (demand?) the existence of scalar stuff different from the pair-valued stuff of our original inquiry. We might only be able to interact with it through the curvature of space, in other words gravity. Is this dark matter? Might we have predicted this possibility ahead of time if we had thought this through 50 years ago?

Author Contributions: All authors have read and agreed to the published version of the manuscript.

Funding: This research received no external funding.

Conflicts of Interest: The authors declare no conflicts of interest.

References

1. Aczél, J. *Lectures on Functional Equations and Their Applications*; Academic Press: New York, NY, USA, 1966.
2. Aczél, J. The associativity equation re-revisited. In Proceedings of the AIP Conference Proceedings, Jackson Hole, WY, USA, 3–8 August 2003; AIP: New York, NY, USA, 2004; Volume 707, pp. 195–203. [CrossRef]
3. Knuth, K.H. Deriving laws from ordering relations. In Proceedings of the AIP Conference Proceedings, Jackson Hole, WY, USA, 3–8 August 2003; AIP: New York, NY, USA, 2004; Volume 707, pp. 204–235.
4. Knuth, K.H.; Skilling, J. Foundations of inference. *Axioms* **2012**, *1*, 38–73. [CrossRef]
5. Knuth, K.H. The deeper roles of mathematics in physical laws. In *Trick or Truth: the Mysterious Connection between Physics and Mathematics*; Aguirre, A., Foster, B., Merali, Z., Eds.; Springer: Berlin/Heidelberg, Germany, 2016; pp. 77–90.
6. Goyal, P.; Knuth, K.H.; Skilling, J. Origin of complex quantum amplitudes and Feynman's rules. *Phys. Rev. A* **2010**, *81*, 022109. [CrossRef]
7. Skilling, J.; Knuth, K.H. The symmetrical foundation of Measure, Probability and Quantum theories. *Ann. Phys.* **2019**, *531*, 1800057. [CrossRef]
8. Skilling, J.; Knuth, K.H. The arithmetic of uncertainty unifies quantum formalism and relativistic spacetime. *arXiv* **2020**, arXiv:2104.05395.
9. Born, M. Zur Quantenmechanik der Stoßvorgänge (Quantum mechanics of collision processes). *Z. Phys.* **1926**, *38*, 803. [CrossRef]
10. Youssef, S. Quantum mechanics as Bayesian complex probability theory. *Mod. Phys. Lett. A* **1994**, *9*, 2571–2586. [CrossRef]
11. Dirac, P.A.M. A new notation for quantum mechanics. In *Mathematical Proceedings of the Cambridge Philosophical Society*; Cambridge University Press: Cambridge, UK, 1939; Volume 35, pp. 416–418.
12. Borrelli, A. Dirac's bra-ket notation and the notion of a quantum state. Styles of Thinking in Science and Technology. In Proceedings of the 3rd International Conference of the European Society for the History of Science, Vienna, Austria, 10–12 September 2008; pp. 361–371.
13. Hurwitz, A. Über die Komposition der quadratischen Formen von beliebig vielen Variablen. In *Mathematische Werke*; Springer: Berlin/Heidelberg, Germany, 1963; pp. 565–571.
14. Baez, J.C. Division algebras and quantum theory. *Found. Phys.* **2012**, *42*, 819–855. [CrossRef]

Proceeding Paper

A Weakly Informative Prior for Resonance Frequencies †

Marnix Van Soom * and Bart de Boer

AI Lab, Vrije Universiteit Brussel, Pleinlaan 2, 1050 Brussels, Belgium; bart@ai.vub.ac.be
* Correspondence: marnix@ai.vub.ac.be
† Presented at the 40th International Workshop on Bayesian Inference and Maximum Entropy Methods in Science and Engineering, online, 4–9 July 2021.

Abstract: We derive a weakly informative prior for a set of ordered resonance frequencies from Jaynes' principle of maximum entropy. The prior facilitates model selection problems in which both the number and the values of the resonance frequencies are unknown. It encodes a weakly inductive bias, provides a reasonable density everywhere, is easily parametrizable, and is easy to sample. We hope that this prior can enable the use of robust evidence-based methods for a new class of problems, even in the presence of multiplets of arbitrary order.

Keywords: weakly uninformative prior; resonance frequency; model selection; maximum entropy

Citation: Van Soom, M.; de Boer, B. A Weakly Informative Prior for Resonance Frequencies. *Phys. Sci. Forum* **2021**, *3*, 2. https://doi.org/10.3390/psf2021003002

Academic Editors: Wolfgang von der Linden and Sascha Ranftl

Published: 4 November 2021

Publisher's Note: MDPI stays neutral with regard to jurisdictional claims in published maps and institutional affiliations.

1. Introduction

An important problem in the natural sciences is the accurate measurement of resonance frequencies. The problem can be formalized by the following probabilistic model:

$$p(D, x|I) = p(D|x)p(x|I) \equiv \mathcal{L}(x)\pi(x), \tag{1}$$

where D is the data, $x = \{x_k\}_{k=1}^{K}$ are the K resonance frequencies of interest, and I is the prior information about x. As an example instance of (1), we refer to the vocal tract resonance (VTR) problem discussed in Section 5 for which D is audio recorded from the mouth of a speaker; x are a set of K VTR frequencies, and the underlying model is a sinusoidal regression model. Any realistic problem will include additional model parameters θ, but these have been silently ignored by formally integrating them out of (1), i.e., $p(D, x|I) = \int d\theta\, p(D, x, \theta|I)$.

In this paper, we assume that the likelihood $\mathcal{L}(x) \equiv p(D|x)$ is given, and our task is to choose an *uninformative* prior $\pi(x) \equiv p(x|I)$ from *limited* prior information I. A conflict arises, however:

> The uninformative priors π most commonly chosen to express limited prior information I are, in practice, often precluded by that same I. $\qquad(2)$

The goal of this paper is to describe this conflict (2) and to show how it can be resolved by adopting a specific choice for π. This allows robust inference of the number of resonances K in the important case of such limited prior information I, which in turn enables accurate measurement of the resonance frequencies x with standard methods such as nested sampling [1] or reversible jump MCMC [2].

2. Notation

The symbol π is intended to convey a vague notion of a generally uninformative or weakly informative prior. Definite *choices* for π are indicated with the subscript i:

$$\pi_i(x) \equiv p(x|\beta_i, I_i), \qquad (i = 1, 2, 3), \tag{3}$$

111

where β_i is a placeholder for the hyperparameter specific to π_i. Note that in the plots below and for the experiments in Section 5, the values of the β_i are always set according to Table 1.

Table 1. The values of the hyperparameters β_i used throughout the paper. All quantities are given in units of Hz.

$k \rightarrow$	0	1	2	3	4	5	6	7	8	9	10
$a = \{a_k\}$		200	600	1400	2900	3500					
$b = \{b_k\}$		1100	3500	4000	4500	5500					
$\overline{x_0} = \{\overline{x_k}\}$	200	500	1000	1500	2000	2500	3000	3500	4000	4500	5000
other				$x_0 = 200$					$x_{\max} = 5500$		

Each π_i uniquely determines a number of important high-level quantities since the likelihood $\mathcal{L}(x)$ and data D are assumed to be given. These quantities are the *evidence* for the model with K resonances:

$$Z_i(K) = \int d^K x\, \mathcal{L}(x) \pi_i(x), \tag{4}$$

the *posterior*:

$$P_i(x) = \frac{\mathcal{L}(x) \pi_i(x)}{Z_i(K)}, \tag{5}$$

and the *information*:

$$H_i(K) = \int d^K x\, P_i(x) \log \frac{P_i(x)}{\pi_i(x)}, \tag{6}$$

which measures the amount of information obtained by updating from prior π_i to posterior P_i, i.e., $H_i(K) \equiv D_{\mathrm{KL}}(P_i|\pi_i)$, where D_{KL} is the Kullback–Leibler divergence.

3. Conflict

The *uninformative priors* π referenced in (2) are of the independent and identically distributed type:

$$\pi(x) = \prod_{k=1}^{K} g(x_k|\beta), \tag{7}$$

where $g(x|\beta)$ is any wide distribution with hyperparameters β. A typical choice for g is the uniform distribution over the full frequency bandwidth; other examples include diffuse Gaussians or Jeffreys priors [3–9].

Second, the *limited prior information I* in (2) about K implies that the problem will involve model selection, since each value of K implicitly corresponds to a different model for the data. It is, thus, necessary to evaluate and compare evidence $Z(K) = \int d^K x\, \mathcal{L}(x) \pi(x)$ for each plausible K.

The conflict between these two elements is due to the *label switching problem*, which is a well-known issue in mixture modeling, e.g., [10]. The likelihood functions $\mathcal{L}(x)$ used in models parametrized by resonance frequencies are typically invariant to switching the label k; i.e., the index k of the frequency x_k has no distinguishable meaning in the model underlying the data. The posterior $P(x) \propto \mathcal{L}(x) \pi(x)$ will inherit this exchange symmetry if the prior is of type (7). Thus, if the model parameters x are well determined by the data D, the posterior landscape will consist of one *primary* mode, which is defined as a mode living in the *ordered region*:

$$\mathcal{R}_K(x_0) = \{x | x_0 \leq x_1 \leq x_2 \leq \cdots \leq x_K\} \quad \text{with} \quad x_0 > 0, \tag{8}$$

and $(K! - 1)$ induced modes, which are identical to the primary mode up to a permutation of the labels k and, thus, live outside of the region $\mathcal{R}_K(x_0)$. The trouble is that correctly taking into account these induced modes during the evaluation of $Z(K)$ requires a surprising amount of extra work in addition to tuning the MCMC method of choice, and that is the label switching problem in our setting. In fact, there is currently no widely accepted solution for the label switching problem in the context of mixture models either [11,12]. This is, then, how in (2) uninformative priors π are "precluded" by the limited information I: the latter implies model selection, which in turn implies evaluating $Z(K)$, which is hampered by the label switching problem due to the exchange symmetry of the former. Therefore, it seems better to try to avoid it by encoding our preference for primary modes directly into the prior. This results in abandoning the uninformative prior π in favor of the weakly informative prior π_3, which is proposed in Section 4 as a solution to the conflict.

We use the VTR problem to briefly illustrate the label switching problem in Figure 1. The likelihood $\mathcal{L}(x)$ is described implicitly in Section 5 and is invariant to switching the labels k because the underlying model function (23) of the regression model is essentially a sum of sinusoids, one for each x_k. As frequencies can be profitably thought of as scale variables ([13], Appendix A), the uninformative prior (7) is represented by

$$\pi_1(x) \equiv p(x|x_0, x_{\max}, I_1) = \prod_{k=1}^{K} h(x_k|x_0, x_{\max}),\tag{9}$$

where $\beta_1 \equiv (x_0, x_{\max})$ are a common lower and upper bound, and

$$h(x|a,b) = \begin{cases} \frac{1}{\log(b/a)} \frac{1}{x} & \text{if } a \leq x \leq b \\ 0 & \text{otherwise} \end{cases} \quad \text{with} \quad \begin{array}{l} a > 0 \\ b < \infty \end{array}\tag{10}$$

is the Jeffreys prior, the conventional uninformative prior for a scale variable [although any prior of the form (7) that is sufficiently uninformative would yield essentially the same results.] We have visualized the posterior landscape $P_1(x)$ in Figure 1 by using the pairwise marginal posteriors $P_1(x_k, x_\ell)$ plotted in blue. Note the exchange symmetry of P_1, which manifests as an (imperfect) reflection symmetry around the dotted diagonal $x_k = x_\ell$ bordering the ordered region $\mathcal{R}_3(x_0)$. The primary mode can be identified by the black dot; all other modes are induced modes. Integrating all $K!$ modes to obtain $Z(K)$ quickly becomes intractable for $Z \gtrsim 4$.

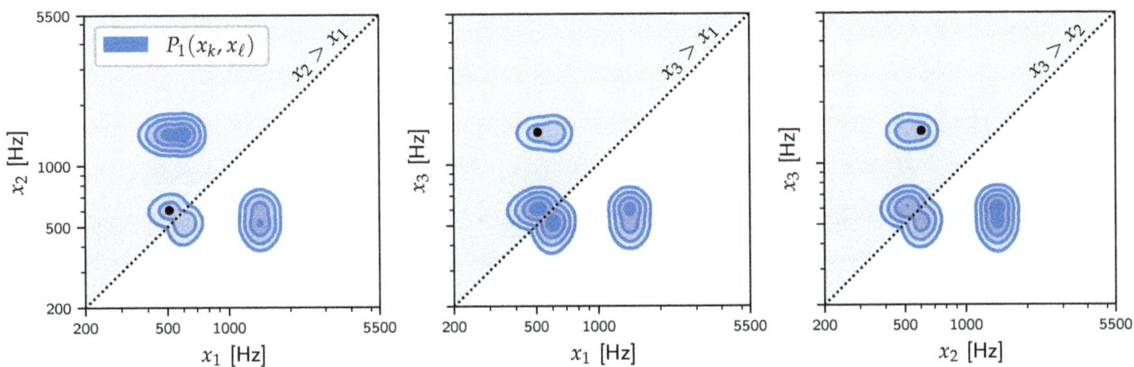

Figure 1. The exchange symmetry of the posterior $P_1(x)$ for a well-determined instance of the VTR problem from Section 5 with $K := 3$. The pairwise marginal posteriors $P_1(x_k, x_\ell)$ are shown using the isocontours of kernel density approximations calculated from posterior samples of x. For each panel, the diagonal $x_k = x_\ell$ is plotted as a dotted line, and the ordered region $\mathcal{R}_3(x_0)$ is shaded in grey. The black dot marks the mean of the primary mode for this problem.

A Simple Way Out?

A simple method out of the conflict is to break the exchange symmetry by assuming specialized bounds for each x_k:

$$\pi_2(x) \equiv p(x|a, b, I_2) = \prod_{k=1}^{K} h(x_k|a_k, b_k), \tag{11}$$

where $\beta_2 \equiv (a, b)$ with $a = \{a_k\}_{k=1}^{K}$ and $b = \{b_k\}_{k=1}^{K}$ being hyperparameters specifying the individual bounds. However, in order to enable the model to detect doublets (a resolved pair of two close frequencies such as the primary mode in the leftmost panel in Figure 1), it is necessary to assign overlapping bounds in (a, b), presumably by using some heuristic. The necessary degree of overlap increases as the detection of higher order multiplets such as triplets (which can and do occur) is desired, but the more overlap in (a, b), the more the label switching problem returns. Despite this issue, there will be cases where we have sufficient prior information I to set the (a, b) hyperparameters without too much trouble; the VTR problem is such a case for which the overlapping values of (a, b) up to $K = 5$ are given in Table 1.

4. Solution

Our solution to the conflict (2) is a chain of K coupled Pareto distributions:

$$\pi_3(x) \equiv p(x|\overline{x_0}, I_3) = \prod_{k=1}^{K} \text{Pareto}(x_k|x_{k-1}, \lambda_k) \tag{12}$$

where

$$\text{Pareto}(x|x_*, \lambda) = \begin{cases} \frac{\lambda x_*^{\lambda}}{x^{\lambda+1}} & \text{if } x \geq x_* \\ 0 & \text{otherwise} \end{cases} \quad \text{with} \quad \begin{matrix} x_* > 0 \\ \lambda > 0, \end{matrix} \tag{13}$$

and the hyperparameter $\beta_3 \equiv \overline{x_0}$ is defined as

$$\overline{x_0} \equiv (\overline{x_0}, \overline{x}), \quad \overline{x_0} := x_0, \quad \overline{x} = \{\overline{x_k}\}_{k=1}^{K}, \quad \lambda_k = \frac{\overline{x_k}}{\overline{x_k} - \overline{x_{k-1}}}. \tag{14}$$

From Figure 2, it can be seen that π_3 encodes weakly informative knowledge about K *ordered frequencies*: (12) and (13) together imply that $\pi_3(x)$ is defined only for $\overline{x} \in \mathcal{R}_K(x_0)$, while nonzero only for $x \in \mathcal{R}_K(x_0)$. In other words, its support is precisely the ordered region $\mathcal{R}_K(x_0)$, which solves the label switching problem underlying the conflict automatically, as the exchange symmetry of π is broken. This is illustrated in Figure 2, where P_3 contracts to a single primary mode, which is just what we would like.

The $K + 1$ hyperparameters $\overline{x_0}$ in (14) are a *common lower bound* x_0 plus K *expected values of the resonance frequencies* $\overline{x}$. While the former is generally easily determined, the latter may seem difficult to set given the premise of this paper that we dispose only of limited prior information I. Why do we claim that π_3 is only weakly informative if it is parametrized by the expected values of the very things it is supposed to be only weakly informative about? The answer is that for any reasonable amount of data, inference based on π_3 is completely insensitive to the exact values of $\overline{x}$. Therefore, any reasonable guess for $\overline{x_0}$ will suffice in practice. For example, for the VTR problem, we simply applied a heuristic where we take $\overline{x_k} = k \times 500$ Hz (see Table 1). This insensitivity is due to the maximum entropy status of π_3 and indicates the weak inductive bias it entails. On a more prosaic level, the heavy tails of the Pareto distributions in (12) ensure that the prior will be eventually overwhelmed by the data no matter how a priori improbable the true value of x is. More prosaic still, in Section 5.1 below we show quantitatively that for the VTR problem π_3 is about as (un)informative as π_2.

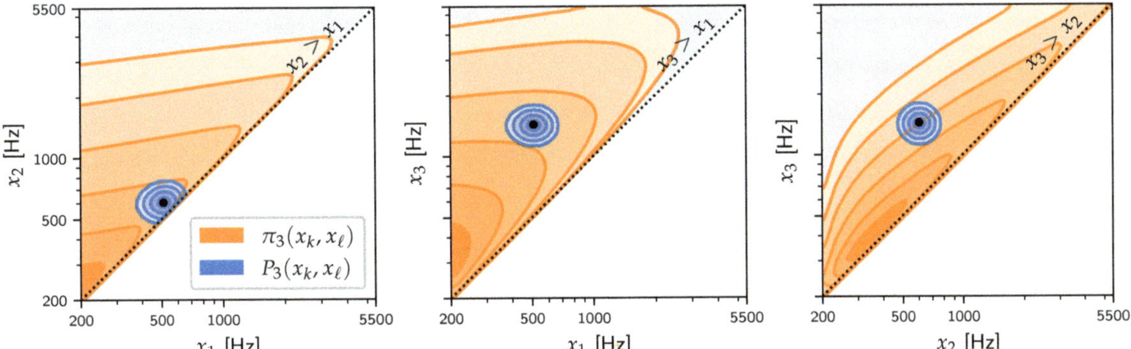

Figure 2. Contraction of prior (π_3) to posterior (P_3) for the application of π_3 to the VTR problem used in Figure 1. The pairwise marginal prior $\pi_3(x_k, x_\ell)$ is obtained by integrating out the third frequency; for example, $\pi_3(x_1, x_2) = \int \mathrm{d}x_3\, \pi_3(x)$. Unlike P_1 in Figure 1, P_3 exhibits only a single mode that coincides with the primary mode as marked by the black dot.

4.1. Derivation of π_3

Our ansatz consists of interpreting the x as a set of K *ordered* scale variables that are bounded from below by x_0. Starting from (9) and not bothering with the bounds (a, b), we obtain the improper pdf

$$m(x) \propto \begin{cases} \prod_{k=1}^{K} \frac{1}{x_k} & x \in \mathcal{R}_K(x_0) \\ 0 & \text{otherwise.} \end{cases} \tag{15}$$

We can simplify (15) using the one-to-one transformation $x \leftrightarrow u$ defined as

$$x \to u: \quad u_k = \log \frac{x_k}{x_{k-1}} \qquad (k = 1, 2, \ldots, K)$$

$$u \to x: \quad x_k = x_0 \exp \sum_{\kappa=1}^{k} u_\kappa \quad (k = 1, 2, \ldots, K) \tag{16}$$

which yields (with abuse of notation for brevity)

$$m(u) \propto \begin{cases} 1 & u \geq 0 \\ 0 & \text{otherwise.} \end{cases} \tag{17}$$

Since model selection requires proper priors, we need to normalize $m(u)$ by adding extra information (i.e., constraints) to it; we propose to simply fix the K first moments $\langle u \rangle = \{\langle u_k \rangle\}_{k=1}^{K}$. This will yield the Pareto chain prior $\pi_3(u)$ directly, expressed in u space rather than x space. The expression for $\pi_3(u)$ is found by minimizing the Kullback–Leibler divergence [14]

$$D_{\mathrm{KL}}(\pi_3 | m) = \int \mathrm{d}^K u\, \pi_3(u) \log \frac{\pi_3(u)}{m(u)}, \quad \text{subject to} \quad \langle u \rangle \equiv \int \mathrm{d}^K u\, u\, \pi_3(u) = \overline{u}, \tag{18}$$

where $\overline{u} = \{\overline{u_k}\}_{k=1}^{K}$ are the supplied first moments. This variational problem is equivalent to finding $\pi_3(u)$ by means of Jaynes' *principle of maximum entropy* with $m(u)$ serving as the invariant measure [15]. Since the exponential distribution $\mathrm{Exp}(x|\lambda)$ is the maximum entropy distribution for a random variable $x \geq 0$ with a fixed first moment $\langle x \rangle = 1/\lambda$, the solution to (18) is

$$\pi_3(u) = \prod_{k=1}^{K} \mathrm{Exp}(u_k | \lambda_k), \tag{19}$$

where the rate hyperparameters $\lambda_k = 1/\overline{u_k}$ and

$$\mathrm{Exp}(x|\lambda) = \begin{cases} \lambda \exp\{-\lambda x\} & \text{if } x \geq 0 \\ 0 & \text{otherwise} \end{cases} \quad \text{with} \quad \lambda > 0. \tag{20}$$

Transforming (19) to x space using (16) finally yields (12), but we still need to express λ_k in terms of $\overline{x}$—we might find it hard to pick reasonable values of $\overline{u_k} = \overline{\log x_k/x_{k-1}}$ from limited prior information I. For this, we will need the identity

$$\langle x_k \rangle \equiv \int \mathrm{d}^K x \, x_k \pi_3(x) = \frac{\lambda_k}{\lambda_k - 1} \langle x_{k-1} \rangle \qquad (k = 1, 2, \ldots, K). \tag{21}$$

Constraining $\langle x_k \rangle = \overline{x_k}$ and solving for λ_k, we obtain $\lambda_k = \overline{x_k}/(\overline{x_k} - \overline{x_{k-1}})$, in agreement with (14). Note that the existence of the first marginal moments $\langle x_k \rangle$ requires that $\lambda_k > 1$.

4.2. Sampling from π_3

Sampling from π_3 is trivial because of the independence of the u_k in u space (19). To produce a sample $x' \sim \pi_3(x)$ given the hyperparameter $\overline{x_0}$, compute the corresponding rate parameters $\{\lambda_k\}_{k=1}^K$ from (14), and use them in (19) to obtain a sample $u' \sim \pi_3(u)$. The desired x' is then obtained from u' using the transformation (16).

5. Application: The VTR Problem

We now present a relatively simple but realistic instance of the problem of measuring resonance frequencies, which will allow us to illustrate the above ideas. The VTR problem consists of measuring human vocal tract resonance (VTR) frequencies x for each of five representative vowel sounds taken from the CMU ARCTIC database [16]. The VTR frequencies x describe the *vocal tract transfer function* $T(x)$ and are fundamental quantities in acoustic phonetics [17]. The five vowel sounds are recorded utterances of the first vowel in the words $W = \{\text{shore}, \text{that}, \text{you}, \text{little}, \text{until}\}$. In order to achieve high-quality VTR frequency estimates $\hat{x}$, only the quasi-periodic steady-state part of the vowel sound is considered for the measurement. The data D, thus, consists of a string of highly correlated *pitch periods*. See Figure 3 for an illustration of these concepts.

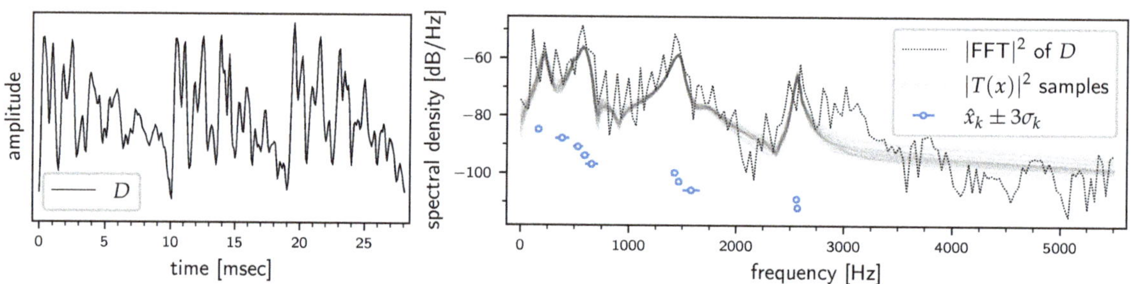

Figure 3. The VTR problem for the case ($D := \text{until}, K := 10$). Left panel: The data D, i.e., the quasi-periodic steady-state part, consist of 3 highly correlated pitch periods. Right panel: Inferred VTR frequency estimates $\{\hat{x}_k\}_{k=1}^K$ for $K := 10$ at 3 sigma. They describe the power spectral density of the vocal tract transfer function $|T(x)|^2$, represented here by 25 posterior samples and compared to the Fast Fourier Transform (FFT) of D. All $\hat{x}_k$ are well resolved, and most have error bars too small to be seen on this scale.

The measurement itself is formalized as inference using the probabilistic model (1). The model assumed to underlie the data is the sinusoidal regression model introduced in [18]; due to limited space, we only describe it implicitly. The sinusoidal regression model assumes that each pitch period $d \in D$ can be modeled as

$$d_t = f(t; A, \alpha, x) + \sigma e_t \quad \text{where} \quad e_t \sim \mathcal{N}(0, 1), \qquad (t = 1, 2, \ldots, T), \qquad (22)$$

where $d = \{d_t\}_{t=1}^T$ is a time series consisting of T samples. The model function

$$f(t; A, \alpha, x) = \sum_{k=1}^{K} [A_k \cos(x_k t) + A_{K+k} \sin(x_k t)] \exp\{-\alpha_k t\} + \sum_{\ell=1}^{L} A_{2K+\ell} t^{\ell-1} \qquad (23)$$

consists of a sinusoidal part (first $\sum$) and a polynomial trend correction (second $\sum$). Note the additional model parameters $\theta = \{A, \alpha, \sigma, L\}$. Formally, given the prior $p(\theta)$ ([18], Section 2.2), the marginal likelihood $\mathcal{L}(x)$ is then obtained as $\mathcal{L}(x) = \int d\theta \, \mathcal{L}(x, \theta) p(\theta)$, where the complete likelihood $\mathcal{L}(x, \theta)$ is implicitly given by (22) and (23). Practically, we just marginalize out θ from samples obtained from the complete problem $p(D, x, \theta | I)$.

For inference, the computational method of choice is nested sampling [1] using the dynesty library [19–23], which scales roughly as $\mathcal{O}(K^2)$ [24]. Since the VTR problem is quite simple ($H_i(K) \sim 30$ nats), we only perform single nested sampling runs and take the obtained $\log Z_i(K)$ and $H_i(K)$ as point estimates. Full details on the experiments and data are available at https://github.com/mvsoom/frequency-prior.

5.1. Experiment I: Comparing π_2 and π_3

In Experiment I, we perform a high-level comparison between π_2 and π_3 in terms of *evidence* (4) and *information* (6). The values of the hyperparameters used in the experiment are listed in Table 1. We did not include π_1 in this comparison as the label switching problem prevented convergence of nested sampling runs for $K \geq 4$. The (a, b) bounds for π_2 were based on loosely interpreting the VTRs as formants and consulting formant tables from standard works [25–30]. These allowed us to compile bounds up until the fifth formant such that $K_{\max} = 5$. For π_3, we simply applied a heuristic where we take $\overline{x_k} = k \times 500$ Hz. We selected x_0 empirically (although a theoretical approach is also possible [31]), and $x_{\max}$ was set to the Nyquist frequency. The role of $x_{\max}$ is to truncate π_3 in order to avoid aliasing effects, since the support of $\pi_3(x_i)$ is unbounded from above. We implemented this by using the following likelihood function in the nested sampling program:

$$\mathcal{L}'(x) = \begin{cases} \mathcal{L}(x) & \text{if } x_k \leq x_{\max} \text{ for all } (k = 1, 2, \ldots, K) \\ 0 & \text{otherwise.} \end{cases} \qquad (24)$$

First, we compare the influence of π_2 and π_3 on model selection. Given $D \in W$, the posterior probability of the number of resonances K is given by the following.

$$p_i(K) = \frac{Z_i(K)}{\sum_{K'} Z_i(K')} \qquad (K = 1, 2, \ldots, K_{\max}). \qquad (25)$$

The results in the top row of Figure 4a are striking: while $p_2(K)$ shows individual preferences based on D, $p_3(K)$ prefers $K = K_{\max}$ unequivocally.

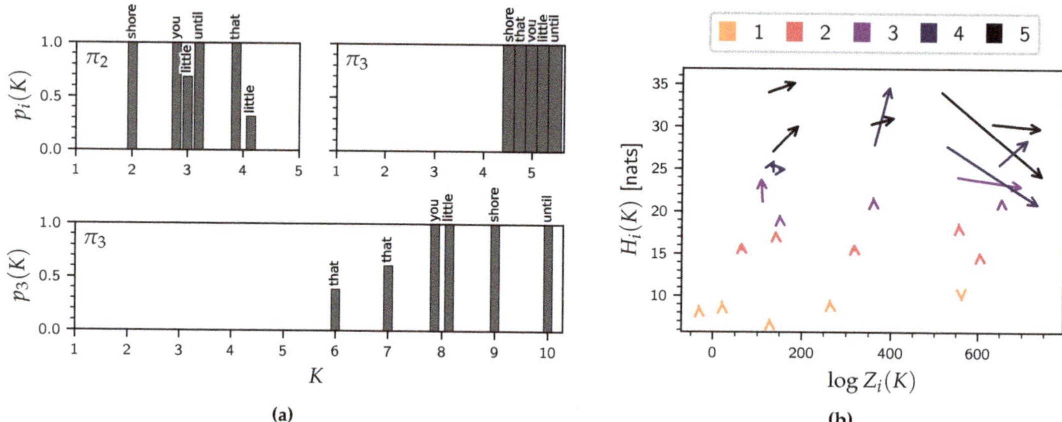

Figure 4. (**a**) Model selection in Experiment I (top row) and Experiment II (bottom row). (**b**) In Experiment I, π_2 and π_3 are compared in terms of evidence [$\log Z_i(K)$] and uninformativeness [$H_i(K)$] for each (D, K). The arrows point from π_2 to π_3 and are color-coded by the value of K. For small values of K, the arrow lengths are too small to be visible on this scale.

Second, in Figure 4b, we compare π_2 and π_3 directly in terms of differences in evidence [$\log Z_i(K)$] and uninformativeness [$H_i(K)$] for each combination (D, K).

Arrows pointing *eastward* indicate $Z_3(K) > Z_2(K)$. The π_3 prior dominates the π_2 prior in terms of evidence, for almost all values of K, indicating that π_3 places its mass in regions of higher likelihood or, equivalently, that the data were much more probable under π_3 than π_2. This implies that the hint of π_3 at more structure beyond $K > K_{\max}$ should be taken serious–we investigate this in Section 5.2.

Arrows pointing *northward* indicate $H_3(K) > H_2(K)$, i.e., π_3 is *less* informative than π_2, since more information is gained by updating from π_3 to P_3 than from π_2 to P_2. It is observed that π_2 and π_3 are roughly comparable in terms of (un)informativeness.

5.2. Experiment II: 'Free' Analysis

We now freely look for more structure in the data by letting K vary up until $K_{\max} = 10$. This goes beyond the capacities of π_1 (because of the label switching problem) and π_2 (because no data are available to set the (a, b) bounds). Thus, the great advantage of π_3 is that we can use a simple heuristic to set $\overline{x_0}$ and let the model perform the discovering without worrying about convergence issues or the obtained evidence values. The bottom row in Figure 4a shows that model selection for the VTR problem is well-defined, with the most probable values of $K \leq 10$, except for $D =$ until. That case is investigated in Figure 3, where the need for more VTRs (higher K) is apparent from the unmodeled broad peak centered at around 3000 Hz in the FFT power spectrum (right panel). Incidentally, this spectrum also shows that spectral peaks are often resolved into more than one VTR, which underlines the importance of using a prior that enables trouble-free handling of multiplets of arbitrary order. A final observation from the spectrum is the fact that the inferred $\hat{x}_k$ differs substantially from the supplied values in $\overline{x}$ (Table 1), which hints at the weak inductive bias underlying π_3.

6. Discussion

It is only when the information in the prior is comparable to the information in the data that the prior probability can make any real difference in parameter estimation problems or in model selection problems ([32], p. 9).

Although the weakly informative prior for resonance frequencies π_3 is meant to be overwhelmed, its practical advantage (i.e., solving the label switching problem) will nonetheless persist, making a real difference in model selection problems even when "the information in the prior" is much smaller than "the information in the data". In this sense, π_3 is quite unlike the prior referenced in the above quote. Since it will be overwhelmed, all it has to do is provide a reasonable density everywhere (which it does), be easily parametrizable (which it is), and be easy to sample from (which it is).

Thus, we hope that this prior can enable the use of robust evidence-based methods for a new class of problems, even in the presence of multiplets of arbitrary order. The prior is compatible with off-the-shelf exploration algorithms and solves the label switching problem without any special tuning or post processing. It would be interesting to compare it to other approaches, e.g., [33], especially in terms of exploration efficiency. It is valid for any collection of scale variables that is intrinsically ordered, of which frequencies and wavelengths seem to be the most natural examples. Some examples of recent work where the prior could be applied directly are:

- Nuclear magnetic resonance (NMR) spectroscopy [34];
- Resonant ultrasound spectroscopy (a standard method in material science) [35];
- In the analysis of atomic spectra [36], such as X-ray diffraction [37];
- Accurate modeling of instrument noise (in this case LIGO/Virgo noise) [38];
- Model-based Bayesian analysis in acoustics [39].

Author Contributions: Conceptualization, writing, methodology, and analysis: M.V.S. Supervision: B.d.B. All authors have read and agreed to the published version of the manuscript.

Funding: This research was funded by the Flemish AI plan and by the Research Foundation Flanders (FWO) under grant number G015617N.

Acknowledgments: We would like to thank Roxana Radulescu, Timo Verstraeten, and Yannick Jadoul for helpful comments.

Conflicts of Interest: The authors declare no conflict of interest.

References

1. Skilling, J. Nested Sampling for General Bayesian Computation. *Bayesian Anal.* **2006**, *1*, 833–859. [CrossRef]
2. Green, P.J. Reversible Jump Markov Chain Monte Carlo Computation and Bayesian Model Determination. *Biometrika* **1995**, *82*, 711–732. [CrossRef]
3. Mark, Y.Z.; Hasegawa-johnson, M. Particle Filtering Approach to Bayesian Formant Tracking. In Proceedings of the IEEE Workshop on Statistical Signal Processing, St. Louis, MO, USA, 28 September–1 October 2003.
4. Zheng, Y.; Hasegawa-Johnson, M. Formant Tracking by Mixture State Particle Filter. In Proceedings of the 2004 IEEE International Conference on Acoustics, Speech, and Signal Processing, Montreal, QC, Canada, 17–21 May 2004; Volume 1, pp. 1–565. [CrossRef]
5. Yan, Q.; Vaseghi, S.; Zavarehei, E.; Milner, B.; Darch, J.; White, P.; Andrianakis, I. Formant Tracking Linear Prediction Model Using HMMs and Kalman Filters for Noisy Speech Processing. *Comput. Speech Lang.* **2007**, *21*, 543–561. [CrossRef]
6. Mehta, D.D.; Rudoy, D.; Wolfe, P.J. Kalman-Based Autoregressive Moving Average Modeling and Inference for Formant and Antiformant Tracking. *J. Acoust. Soc. Am.* **2012**, *132*, 1732–1746. [CrossRef]
7. Shi, Y.; Chang, E. Spectrogram-Based Formant Tracking via Particle Filters. In Proceedings of the (ICASSP '03), 2003 IEEE International Conference on Acoustics, Speech, and Signal Processing, Hong Kong, China, 6–10 April 2003; Volume 1, p. 1. [CrossRef]
8. Deng, L.; Lee, L.J.; Attias, H.; Acero, A. Adaptive Kalman Filtering and Smoothing for Tracking Vocal Tract Resonances Using a Continuous-Valued Hidden Dynamic Model. *IEEE Trans. Audio Speech Lang. Process.* **2007**, *15*, 13–23. [CrossRef]
9. Luberadzka, J.; Kayser, H.; Hohmann, V. Glimpsed Periodicity Features and Recursive Bayesian Estimation for Modeling Attentive Voice Tracking. *Int. Congr. Acoust.* **2019**, *9*, 8.
10. Stephens, M. Dealing with Label Switching in Mixture Models. *J. R. Stat. Soc. Ser. (Stat. Methodol.)* **2000**, *62*, 795–809. [CrossRef]
11. Celeux, G.; Kamary, K.; Malsiner-Walli, G.; Marin, J.M.; Robert, C.P. Computational Solutions for Bayesian Inference in Mixture Models. *arXiv* **2018**, arXiv:1812.07240.
12. Celeux, G.; Fruewirth-Schnatter, S.; Robert, C.P. Model Selection for Mixture Models - Perspectives and Strategies. *arXiv* **2018**, arXiv:1812.09885.
13. Bretthorst, G.L. *Bayesian Spectrum Analysis and Parameter Estimation*; Springer: Berlin/Heidelberg, Germany, 1988.
14. Knuth, K.H.; Skilling, J. Foundations of Inference. *Axioms* **2012**, *1*, 38–73. [CrossRef]
15. Jaynes, E.T. Prior Probabilities. *IEEE Trans. Syst. Sci. Cybern.* **1968**, *4*, 227–241. [CrossRef]
16. Kominek, J.; Black, A.W. The CMU Arctic Speech Databases. In Proceedings of the Fifth ISCA Workshop on Speech Synthesis, Pittsburgh, PA, USA, 14–16 June 2004.

17. Van Soom, M.; de Boer, B. A New Approach to the Formant Measuring Problem. *Proceedings* **2019**, *33*, 29. [CrossRef]
18. Van Soom, M.; de Boer, B. Detrending the Waveforms of Steady-State Vowels. *Entropy* **2020**, *22*, 331. [CrossRef] [PubMed]
19. Speagle, J.S. Dynesty: A Dynamic Nested Sampling Package for Estimating Bayesian Posteriors and Evidences. *arXiv* **2019**, arXiv:1904.02180.
20. Feroz, F.; Hobson, M.P.; Bridges, M. MULTINEST: An Efficient and Robust Bayesian Inference Tool for Cosmology and Particle Physics. *Mon. Not. R. Astron. Soc.* **2009**, *398*, 1601–1614. [CrossRef]
21. Neal, R.M. Slice Sampling. *Ann. Stat.* **2003**, *31*, 705–767. [CrossRef]
22. Handley, W.J.; Hobson, M.P.; Lasenby, A.N. POLYCHORD: Nested Sampling for Cosmology. *Mon. Not. R. Astron. Soc.* **2015**, *450*, L61–L65. [CrossRef]
23. Handley, W.J.; Hobson, M.P.; Lasenby, A.N. POLYCHORD: Next-Generation Nested Sampling. *Mon. Not. R. Astron. Soc.* **2015**, *453*, 4384–4398. [CrossRef]
24. Buchner, J. Nested Sampling Methods. *arXiv* **2021**, arXiv:2101.09675.
25. Peterson, G.E.; Barney, H.L. Control Methods Used in a Study of the Vowels. *J. Acoust. Soc. Am.* **1952**, *24*, 175–184. [CrossRef]
26. Hillenbrand, J.; Getty, L.A.; Clark, M.J.; Wheeler, K. Acoustic Characteristics of American English Vowels. *J. Acoust. Soc. Am.* **1995**, *97*, 3099–3111. [CrossRef]
27. Vallée, N. Systèmes Vocaliques: De La Typologie Aux Prédictions, Ph.D. Thesis, Université Stendhal, Grenoble, France, 1994.
28. Kent, R.D.; Vorperian, H.K. Static Measurements of Vowel Formant Frequencies and Bandwidths: A Review. *J. Commun. Disord.* **2018**, *74*, 74–97. [CrossRef]
29. Vorperian, H.K.; Kent, R.D.; Lee, Y.; Bolt, D.M. Corner Vowels in Males and Females Ages 4 to 20 Years: Fundamental and F1–F4 Formant Frequencies. *J. Acoust. Soc. Am.* **2019**, *146*, 3255–3274. [CrossRef] [PubMed]
30. Klatt, D.H. Software for a Cascade/Parallel Formant Synthesizer. *J. Acoust. Soc. Am.* **1980**, *67*, 971–995. [CrossRef]
31. de Boer, B. Acoustic Tubes with Maximal and Minimal Resonance Frequencies. *J. Acoust. Soc. Am.* **2008**, *123*, 3732. [CrossRef]
32. Bretthorst, G.L. Bayesian Analysis. II. Signal Detection and Model Selection. *J. Magn. Reson.* **1990**, *88*, 552–570. [CrossRef]
33. Buscicchio, R.; Roebber, E.; Goldstein, J.M.; Moore, C.J. Label Switching Problem in Bayesian Analysis for Gravitational Wave Astronomy. *Phys. Rev. D* **2019**, *100*, 084041. [CrossRef]
34. Wilson, A.G.; Wu, Y.; Holland, D.J.; Nowozin, S.; Mantle, M.D.; Gladden, L.F.; Blake, A. Bayesian Inference for NMR Spectroscopy with Applications to Chemical Quantification. *arXiv* **2014**, arXiv:1402.3580.
35. Xu, K.; Marrelec, G.; Bernard, S.; Grimal, Q. Lorentzian-Model-Based Bayesian Analysis for Automated Estimation of Attenuated Resonance Spectrum. *IEEE Trans. Signal Process.* **2019**, *67*, 4–16. [CrossRef]
36. Trassinelli, M. Bayesian Data Analysis Tools for Atomic Physics. *Nucl. Instruments Methods Phys. Res. Sect. Beam Interact. Mater. Atoms* **2017**, *408*, 301–312. [CrossRef]
37. Fancher, C.M.; Han, Z.; Levin, I.; Page, K.; Reich, B.J.; Smith, R.C.; Wilson, A.G.; Jones, J.L. Use of Bayesian Inference in Crystallographic Structure Refinement via Full Diffraction Profile Analysis. *Sci. Rep.* **2016**, *6*, 31625. [CrossRef] [PubMed]
38. Littenberg, T.B.; Cornish, N.J. Bayesian Inference for Spectral Estimation of Gravitational Wave Detector Noise. *Phys. Rev. D* **2015**, *91*, 084034. [CrossRef]
39. Xiang, N. Model-Based Bayesian Analysis in Acoustics-A Tutorial. *J. Acoust. Soc. Am.* **2020**. [CrossRef] [PubMed]

physical sciences forum

MDPI

Proceeding Paper

An Entropic Approach to Classical Density Functional Theory [†]

Ahmad Yousefi * and Ariel Caticha

Physics Department, University at Albany-SUNY, Albany, NY 12222, USA; acaticha@albany.edu

* Correspondence: ahmad369@gmail or ayousefi@albany.edu

† Presented at the 40th International Workshop on Bayesian Inference and Maximum Entropy Methods in Science and Engineering, online, 4–9 July 2021.

Abstract: The classical Density Functional Theory (DFT) is introduced as an application of entropic inference for inhomogeneous fluids in thermal equilibrium. It is shown that entropic inference reproduces the variational principle of DFT when information about the expected density of particles is imposed. This process introduces a family of trial density-parametrized probability distributions and, consequently, a trial entropy from which the preferred one is found using the method of Maximum Entropy (MaxEnt). As an application, the DFT model for slowly varying density is provided, and its approximation scheme is discussed.

Keywords: entropic inference; relative entropy; density functional theory; contact geometry; optimal approximations

1. Introduction

The Density Functional Theory was first developed in the context of quantum mechanics and only later extended to the classical regime. The theory was first introduced by Kohn and Hohenberg (1964) [1] as a computational tool to calculate the spatial density of an electron gas in the presence of an external potential at zero temperature. Soon afterwards, Mermin provided the extension to finite temperatures [2]. Ebner, Saam, and Stroud (1976) [3] applied the idea to simple classical fluids, and Evans (1979) provided a systematic formulation in his classic paper [4]: "The nature of the liquid–vapour interface and other topics in the statistical mechanics of non-uniform, classical fluids".

The majority of physicists and chemists today are aware of the quantum DFT and the Kohn–Sham model [5], while fewer are familiar with the classical DFT; a historical review of quantum DFT and its vast variety of applications is found in [6,7]. The classical DFT, similarly, is a "formalism designed to tackle the statistical mechanics of inhomogeneous fluids" [8], which has been used to investigate a wide variety of equilibrium phenomena, including surface tension, adsorption, wetting, fluids in porous materials, and the chemical physics of solvation.

Just as the Thomas–Fermi–Dirac theory is usually regarded as a precursor of quantum DFT, van der Waals' thermodynamic theory of capillarity under the hypothesis of a continuous variation of density [9] can be regarded as the earliest work on classical DFT without a fundamental proof of existence for such a variational principle.

"The long-term legacy of DFT depends largely on the continued value of the DFT computer programs that practitioners use daily" [6]. The algorithms behind the computer programs, all starting from an original Hartree–Fock method to solve the N-particle Schrödinger equation, have evolved with many approximations and extensions implemented over time by a series of individuals; although the algorithms produce accurate results, they do not mention the HK variational principle. Without the variational principle, the computer codes are suspected of being ad hoc or intuitively motivated without a solid theoretical foundation; therefore, the DFT variational principle not only scientifically justifies the DFT algorithms, but it also provides us with a basis to understand the repeatedly modified algorithms behind the codes.

Citation: Yousefi, A.; Caticha, A. An Entropic Approach to Classical Density Functional Theory. *Phys. Sci. Forum* **2021**, *3*, 13. https://doi.org/10.3390/psf2021003013

Academic Editors: Wolfgang von der Linden and Sascha Ranftl

Published: 24 December 2021

Publisher's Note: MDPI stays neutral with regard to jurisdictional claims in published maps and institutional affiliations.

In this work, we derive the classical DFT as an application of the method of maximum entropy [10–14]. This integrates the classical DFT with other formalisms of classical statistical mechanics (canonical, grand canonical, etc.) as an application of information theory. Our approach not only enables one to understand the theory from the Bayesian point of view, but also provides a framework to construct equilibrium theories on the foundation of MaxEnt. We emphasize that our goal is not derive an alternative to DFT. Our goal is purely conceptual. We wish to find a new justification or derivation of DFT that makes it explicit how DFT fits within the MaxEnt approach to statistical mechanics. The advantage of such an understanding is the potential for future applications that are outside the reach of the current version of DFT.

In Section 2, we briefly review entropic inference as an inference tool that updates probabilities as degrees of rational belief in response to new information. Then, we show that any entropy maximization produces a contact structure that is invariant under the Legendre Transformations; this enables us to take advantage of these transformations for maximized entropy functions (functionals here) found from constraints other than those of thermal equilibrium, as well as thermodynamic potentials.

In Section 3, we briefly review the method of relative entropy for optimal approximation of probabilities, which allows us to derive and then generalize the Bogolyubov variational principle. Then, we apply it for the special case wherein the trial family of probabilities are parametrized by the density function $n(x)$.

In Section 4, the Density Functional formalism is introduced as an extension of the existing ensemble formalisms of statistical mechanics (canonical, grand canonical, etc.), and we show that the core DFT theorem is an immediate consequence of MaxEnt; we prove that in the presence of an external potential $v(x)$, there exists a trial density functional entropy $S_v(E; n]$ maximized at the equilibrium density. We also prove that this entropy maximization is equivalent to minimization of a density functional potential $\Omega(\beta; n]$ given by

$$\Omega_v(\beta; n] = \int d^3x v(x)n(x) + F(\beta; n] \tag{1}$$

where $F(\beta; n]$ is independent of $v(x)$. This formulation achieves two objectives. **(i)** It shows that the density functional variational principle is an application of MaxEnt for non-uniform fluids at equilibrium, and therefore, varying the density $n(x)$ in $S_v(E; n]$ **does not** imply that the functional represents entropy of any non-equilibrium system. This trial entropy, although very useful, is just a mathematical construct that allows us to incorporate constraints that are related to one another by definition. **(ii)** By this approach, we show that the Bayesian interpretation of probability liberates the fundamental theorem of the DFT from an imaginary grand-canonical ensemble, i.e., the thermodynamic chemical potential is appropriately defined without the need to define microstates for varying numbers of particles.

Finally, in Section 5, as an illustration, we discuss the already well-known example of a slowly varying inhomogeneous fluid. We show that our entropic DFT allows us to reproduce the gradient approximation results derived by Evans [4]. There are two different approximations involved: (i) rewriting the non-uniform direct correlation function in terms of the uniform one and (ii) the use of linear response theory to evaluate the Fourier transform of direct correlation functions. The former assumes that the density is uniform inside each volume element, and the latter assumes that in of densities for neighboring volume elements is small compared to their average.

2. Entropic Inference

A discussion of the method of maximum entropy as a tool for inference is found in [14]. Given a **prior** Probability Distribution Function (**PDF**) $Q(X)$, we want to find the **posterior** PDF $P(X)$ subject to constraints on expected values of functions of X.

Formally, we need to maximize the relative entropy

$$S_r[P|Q] \equiv - \sum_X (P \log P - P \log Q) , \tag{2}$$

under constraints $A_i = \sum_X P(X) \hat{A}_i(X)$ and $1 = \sum_X P(X)$, where A_is are real numbers, $\hat{A}_i$s are real-valued functions on the space of X, and $1 \leq i \leq m$ for m number of constraints.

The maximization process yields the posterior probability

$$P(X) = Q(X) \frac{1}{Z} e^{-\Sigma_i \alpha_i \hat{A}_i(X)} , \qquad \text{where } Z = \sum_X Q(X) e^{-\Sigma_i \alpha_i \hat{A}_i(X)} , \tag{3}$$

and α_is are Lagrange multipliers associated with A_is.

Consequently, the maximized entropy is

$$S = \sum_i \alpha_i A_i + \log Z . \tag{4}$$

Now we can show that the above entropy readily produces a contact structure, we can calculate the complete differential of Equation (4) to define the vanishing one-form ω_{cl} as

$$\omega_{cl} \equiv dS - \sum_i \alpha_i dA_i = 0 . \tag{5}$$

Therefore, any classical entropy maximization with m constraints produces a contact structure $\{\mathbb{T}, \omega_{cl}\}$ in which manifold $\mathbb{T}$ has $2m + 1$ coordinates $\{q_0, q_1, \ldots q_m, p_1, \ldots, p_m\}$.

The physically relevant manifold $\mathbb{M}$ is an m-dimensional sub-manifold of $\mathbb{T}$ on which ω_{cl} vanishes; i.e., $\mathbb{M}$ is determined by $1 + m$ equations:

$$q_0 \equiv S(\{A_i\}) , \qquad p_i \equiv \alpha_i = \frac{\partial S}{\partial A_i} . \tag{6}$$

The Legendre Transformations, which are defined as

$$q_0 \longrightarrow q_0 - \sum_{j=1}^{l} p_j q_j , \tag{7}$$

$$q_i \longrightarrow p_i , \quad p_i \longrightarrow -q_i , \qquad \text{for } 1 \leq i \leq l ,$$

are coordinate transformations on space $\mathbb{T}$ under which ω_{cl} is conserved. It has been shown [15,16] that the laws of thermodynamics produce a contact structure conforming to the above prescription. Here, we are emphasizing that the contact structure is an immediate consequence of MaxEnt, and therefore, it can be utilized in applications of information theory beyond thermodynamics.

3. MaxEnt and Optimal Approximation of Probabilities

The posterior PDF found from entropic inference is usually too complicated to be used for practical purposes. A common solution is to approximate the posterior PDF with a more tractable family of PDFs $\{p_\theta\}$ [17]. Given the exact probability p_0, the preferred member of tractable family p_{θ^*} is found by maximizing the entropy of p_θ relative to p_0:

$$\left. \frac{\delta S_r[p_\theta | p_0]}{\delta \theta} \right|_{\theta = \theta^*} = 0 . \tag{8}$$

The density functional formalism is a systematic method in which the family of trial probabilities is parametrized by the density of particles; in Section 4, we shall use the method of maximum entropy to determine the family of trial distributions parametrized by $n(x)$, $p_\theta \equiv p_n$. So, we can rewrite Equation (8) as

$$\frac{\delta}{\delta n(x')}\left[S_r[p_n|p_0] + \alpha_{eq}[N - \int d^3x n(x)]\right]\Bigg|_{n(x)=n_{eq}(x)} = 0 \,. \tag{9}$$

We will see that the canonical distribution itself is a member of the trial family; therefore, in this case, the exact solution to Equation (9) is p_0 itself:

$$p_n\Big|_{n(x)=n_{eq}(x)} = p_0 \,. \tag{10}$$

4. Density Functional Formalism

An equilibrium formalism of statistical mechanics is a relative entropy maximization process consisting of three crucial elements: (i) One must choose the microstates that describe the system of inference. (ii) The prior is chosen to be uniform. (iii) One must select the constraints that represent the information that is relevant to the problem at hand.

In the density ensemble, microstates of the system are given as positions and momenta of all N particles of the same kind, given the uniform prior probability distribution

$$Q(\{\vec{x}_1,\ldots,\vec{x}_N; \vec{p}_1,\ldots,\vec{p}_N\}) = constant. \tag{11}$$

Keeping in mind that we are looking for **thermal properties of inhomogeneous fluids**, it is natural to choose the density of particles $n(x)$ as a computational constraint and the expected energy E as a thermodynamic constraint, in which $n(x)$ represents the **inhomogeneity** and E defines the thermal equilibrium.

Note that all constraints (computational, thermal, etc.) in the framework can be incorporated as inferential constraints and can be imposed as prescribed in Section 2.

The density constraint holds for every point in space; therefore, we have $1 + 1 + \mathbb{R}^3$ constraints—one for normalization, one for total energy, and one for density of particles at each point in space; thus, we have to maximize the relative entropy

$$S_r[P|Q] = -\frac{1}{N!}\int d^{3N}x d^{3N}p (Plog P - Plog Q) \equiv -Tr_c(Plog P - Plog Q), \tag{12}$$

subject to constraints

$$1 = \langle 1 \rangle, \quad E = \langle \hat{H}_v \rangle, \tag{13a}$$

$$n(x) = \langle \hat{n}_x \rangle \quad \text{where} \int d^3x n(x) = N, \tag{13b}$$

where $\langle . \rangle \equiv \frac{1}{N!}\int(.)Pd^{3N}x d^{3N}p$. The classical Hamiltonian operator $\hat{H}$ and the particle density operator $\hat{n}_x$ are given as

$$\hat{H}_v \equiv \sum_{i=1}^{N} v(x_i) + \hat{K}(p_1,\ldots,p_N) + \hat{U}(x_1,\ldots,x_N), \tag{14}$$

$$\hat{n}_x \equiv \sum_{i}^{N} \delta(x - x_i). \tag{15}$$

The density $n(x)$ is not an arbitrary function; it is constrained by a fixed total number of particles,

$$\int d^3x n(x) = N. \tag{16}$$

Maximizing (12) subject to (13) gives the posterior probability $P(x_1, \ldots, x_N; p_1, \ldots, p_N)$ as

$$P = \frac{1}{Z_v} e^{-\beta \hat{H}_v - \int d^3x \alpha(x) \hat{n}_x} \quad \text{under condition} \int d^3x\, n(x) = N. \tag{17}$$

where $\alpha(x)$ and β are Lagrange multipliers.

The Lagrange multiplier function $\alpha(x)$ is implicitly determined by

$$\frac{\delta log Z_v}{\delta \alpha(x)} = -n(x) \tag{18}$$

and by Equation (16)

$$-\int d^3x \frac{\delta log Z_v}{\delta \alpha(x)} = N . \tag{19}$$

Substituting the trial probabilities from (17) into (12) gives the **trial** entropy $S_v(E; n]$ as

$$S_v(E; n] = \beta E + \int d^3x \alpha(x) n(x) + log Z_v, \tag{20}$$

where $Z_v(\beta; \alpha]$ is the **trial** partition function defined as

$$Z_v(\beta; \alpha] = Tr_c e^{-\beta \hat{H}_v - \int d^3x \alpha(x) \hat{n}_x} . \tag{21}$$

The equilibrium density $n_{eq}(x)$ is that which maximizes $S_v(E; n]$ subject to $\int d^3x\, n(x) = N$:

$$\frac{\delta}{\delta n(x')} \left[S_v(E; n] + \alpha_{eq}[N - \int d^3x\, n(x)] \right] = 0 \quad \text{for fixed E.} \tag{22}$$

Next, perform a Legendre transformation and define the Massieu functional $\tilde{S}_v(\beta, n]$ as

$$\tilde{S}_v \equiv S_v - \beta E, \tag{23}$$

so that we can rewrite Equation (22) as

$$\frac{\delta}{\delta n(x')} \left[\tilde{S}_v(\beta; n] - \alpha_{eq} \int d^3x\, n(x) \right] = 0 \quad \text{for fixed } \beta . \tag{24}$$

Combine (20), (23), and (24) and use the variational derivative identity $\frac{\delta n(x)}{\delta n(x')} = \delta(x - x')$ to find

$$\int d^3x \left[\frac{\delta log Z_v(\beta; \alpha]}{\delta \alpha(x)} + n(x) \right] \frac{\delta \alpha(x)}{\delta n(x')} = \alpha_{eq} - \alpha(x'). \tag{25}$$

The LHS of equation (25) vanishes by (16), and therefore, the RHS must vanish for an arbitrary $n(x)$, which implies that

$$\alpha(x) = \alpha_{eq} , \quad \text{and} \quad \left. \frac{\delta log Z_v}{\delta \alpha(x)} \right|_{\alpha_{eq}} = -n_{eq}(x) . \tag{26}$$

Substituting (26) into (17) yields the equilibrium probability distribution

$$P^*(x_1, \ldots, x_N; p_1, \ldots, p_N) = \frac{1}{Z_v^*} e^{-\beta \hat{H}_v - \alpha_{eq} \int d^3x \hat{n}_x} = \frac{1}{Z_v^*} e^{-\beta \hat{H}_v - \alpha_{eq} N} \tag{27}$$

where $Z_v^*(\beta, \alpha_{eq}) = Tr_c e^{-\beta \hat{H}_v - \alpha_{eq} N}$.

From the inferential point of view, **the variational principle for the grand potential and the equilibrium density** [4] is proved at this point; we showed that for an arbitrary classical Hamiltonian $\hat{H}_v$, there exists a trial entropy $S_v(E; n(x)]$ defined by Equation (20), which assumes its maximum at fixed energy and varying $n(x)$ under the condition $\int d^3x\, n(x) = N$ at the equilibrium density and gives the posterior PDF (Equation (27)) equal to that of the canonical distribution.

The massieu function $\tilde{S}_v(\beta; n(x)]$ from Equation (23) defines the **density functional potential** $\Omega_v(\beta; n(x)]$ by

$$\Omega_v(\beta; n] \equiv \frac{-\tilde{S}_v(\beta; n]}{\beta} = -\int d^3x \frac{\alpha(x)}{\beta} n(x) - \frac{1}{\beta} log Z_v(\beta; \alpha] \,, \tag{28}$$

so that the maximization of $S_v(E; n]$ (20) in the vicinity of the equilibrium is equivalent to the minimization of $\Omega_v(\beta; n(x)]$ (28) around the same equilibrium point

$$\frac{\delta}{\delta n(x')}\left[\Omega_v(\beta; n] + \frac{\alpha_{eq}}{\beta}\int d^3x\, n(x)\right] = 0 \,. \tag{29}$$

After we find Ω_v, we just need to recall that $\alpha(x) = -\beta\frac{\delta\Omega_v}{\delta n(x)}$ and substitute in Equation (26) to recover the "core integro-differential equation" [4] of the DFT as

$$\nabla\left(\frac{\delta\Omega(\beta; n]}{\delta n(x)}\right)_{eq} = 0 \tag{30}$$

which implies that

$$\Omega_{v;eq} \leq \Omega_v(\beta; n], \tag{31}$$

where

$$\Omega_{v,eq}(\beta; n] = -\frac{\alpha_{eq}}{\beta}\int d^3x\, n(x) - \frac{1}{\beta} log Z_v^*(\beta, \alpha_{eq}). \tag{32}$$

From Equation (28), it is clear that

$$\Omega_v(\beta; n] = \int d^3x\, v(x)n(x) + \langle \hat{K} + \hat{U}\rangle - \frac{S_v(E; n]}{\beta}\,. \tag{33}$$

It is convenient to define the **intrinsic density functional potential** F_v as

$$F_v(\beta; n] \equiv \langle \hat{K} + \hat{U}\rangle - \frac{S_v}{\beta} \tag{34}$$

to have

$$\Omega_v(\beta; n] = \int v(x)n(x) + F_v(\beta; n] \,. \tag{35}$$

Now we are ready to restate the fundamental theorem of the classical DFT:

Theorem 1. *The intrinsic functional potential F_v is a functional of density $n(x)$ and is independent of the external potential:*

$$\frac{\delta F_v(\beta; n]}{\delta v(x')} = 0 \quad \text{for fixed } \beta \text{ and } n(x). \tag{36}$$

Proof. The crucial observation behind the DFT formalism is that P and Z_v depend on the external potential $v(x)$ and the Lagrange multipliers $\alpha(x)$ only through the particular combination $\bar{\alpha}(x) \equiv \beta v(x) + \alpha(x)$. Substitute Equation (20) into (34) to get

$$\beta F_v(\beta; n] = log Z(\beta; \bar{\alpha}] + \int d^3x\, \bar{\alpha}(x)n(x), \tag{37}$$

where $Z(\beta; \bar{\alpha}] = Z_v(\beta; \alpha] = Tr_c e^{-\beta(\hat{K}+\hat{U}) - \int d^3x\bar{\alpha}(x)\hat{n}_x}$. The functional derivative of βF_v at fixed $n(x)$ and β is

$$\left.\frac{\delta(\beta F_v(\beta; n])}{\delta v(x')}\right|_{\beta, n(x)} = \int d^3x'' \frac{\delta}{\delta\bar{\alpha}(x'')}\left[logZ(\beta; \bar{\alpha}] + \int d^3x\bar{\alpha}(x)n(x)\right]\left.\frac{\delta\bar{\alpha}(x'')}{\delta v(x')}\right|_{\beta, n(x)} . \tag{38}$$

Since $n(x') = -\frac{\delta logZ(\beta;\bar{\alpha}]}{\delta\bar{\alpha}(x')}$, keeping $n(x)$ fixed is achieved by keeping $\bar{\alpha}(x)$ fixed:

$$\left.\frac{\delta\bar{\alpha}(x'')}{\delta v(x')}\right|_{\beta, n(x)} = \left.\frac{\delta\bar{\alpha}(x'')}{\delta v(x')}\right|_{\beta, \bar{\alpha}(x)} = 0 , \tag{39}$$

so that

$$\left.\frac{\delta F_v(\beta, n]}{\delta v(x')}\right|_{\beta, n(x)} = 0 , \tag{40}$$

which concludes the proof; thus, we can write down the intrinsic potential as

$$F(\beta, n(x)] = F_v(\beta, n(x)] . \tag{41}$$

$\square$

Remark 1. *Note that since a change in the external potential $v(x)$ can be compensated by a suitable change in the multiplier $\alpha(x)$ in such a way as to keep $\bar{\alpha}(x)$ fixed, such changes in $v(x)$ will have no effect on $n(x)$. Therefore, keeping $n(x)$ fixed on the left-hand side of (37) means that $\bar{\alpha}(x)$ on the right side is fixed too.*

Now, we can substitute Equation (35) into (29) and define the chemical potential

$$\mu \equiv \frac{-\alpha_{eq}}{\beta} , \tag{42}$$

to have

$$\left.\frac{\delta}{\delta n(x')}\left[\int d^3xv(x)n(x) + F(\beta; n] - \mu\int d^3xn(x)\right]\right|_{n(x)=n_{eq}(x)} = 0 . \tag{43}$$

We can also substitute (35) into (30) to find

$$v(x) + \left.\frac{\delta F}{\delta n(x)}\right|_{eq} = \mu, \tag{44}$$

which allows us to define and **interpret** $\mu_{in}(x; n] \equiv \frac{\delta F}{\delta n(x)}$ as the **intrinsic chemical potential** of the system. To proceed further, we also split F into that of ideal gas plus the interaction part as

$$F(\beta; n] = F_{id}(\beta; n] - \phi(\beta; n]. \tag{45}$$

Differentiating with $\frac{\delta}{\delta n(x)}$ gives

$$\beta\mu_{in}(x; n] = log(\lambda^3 n(x)) - c(x; n] , \tag{46}$$

where, for a monatomic gas, $\lambda = \left(\frac{2\pi\hbar^2}{m}\right)^{1/2}$. The additional one-body potential $c(x; n] = \frac{\delta\phi}{\delta n(x)}$ is related to the Ornstein–Zernike direct correlation function of non-uniform fluid [18–20] by

$$c^{(2)}(x, x'; n] \equiv \frac{\delta c(x; n]}{\delta n(x')} = \frac{\delta^2\phi(\beta; n]}{\delta n(x)\delta n(x')} . \tag{47}$$

5. Slowly Varying Density and Gradient Expansion

We have proved that the solution to Equation (43) is the equilibrium density. However, the functional $F(\beta; n]$ needs to be approximated because the direct calculation of F involves calculating the canonical partition function, the task that we have been avoiding to begin with. Therefore, different models of the DFT may vary in their approach to guessing $F(\beta; n]$. Now assume that we are interested in a monatomic fluid with a slowly varying external potential. In our language, it means that we use the approximation $\int d^3x \equiv \sum (\Delta x)^3$, where Δx is much longer than the density correlation length, and the change in density in each volume element is small compared to the average density. This allows us to interpret each volume element $(\Delta x)^3$ as a fluid at grand canonical equilibrium with the rest of the fluid as its thermal and particle bath.

Similarly to [4], we expand $F(\beta; n]$ as

$$F(\beta; n] = \int d^3x \left[f_0(n(x)) + f_2(n(x))|\nabla n(x)|^2 + \mathcal{O}(\nabla^4 n(x)) \right]. \tag{48}$$

Differentiating with respect to $n(x)$, we have

$$\mu_{in}(x; n] = \frac{\delta F}{\delta n(x)} = f_0'(n(x)) - f_2'(n(x))|\nabla n(x)|^2 - 2f_2(n(x))\nabla^2 n(x). \tag{49}$$

In the absence of an external potential, $v(x) = 0$, the second and the third terms in the RHS of (49) vanish, and from Equation (44), $\mu_{in} = \mu$; therefore, we have

$$f_0'(n) = \mu(n(x)), \tag{50}$$

where $\mu(n(x))$ is the chemical potential of a uniform fluid with density $n = n(x)$. On the other hand, with the assumption that each volume element behaves as if it is in grand canonical ensemble for itself under influence of both external potential and additional one-body interaction $c(x; n]$, we know that the second derivative of F is related to Ornstein–Zernike theory by

$$\beta \frac{\delta^2 F}{\delta n(x)\delta n(x')} = \frac{\delta(x - x')}{n(x)} - c^{(2)}(x, x'; n]. \tag{51}$$

Therefore we have a Taylor expansion of F around the uniform density as

$$F[n(x)] = F[n] + \int d^3x \left[\frac{\delta F}{\delta n(x)}\right]_{n_{eq}(x)} \tilde{n}(x) \tag{52}$$

$$+ \frac{1}{2\beta} \int \int d^3x d^3x' \left[\frac{\delta(x - x')}{n(x)} - c^{(2)}(|x - x'|; n]\right]_{n_{eq}(x)} \tilde{n}(x)\tilde{n}(x') + \dots,$$

where $\tilde{n}(x) \equiv n(x) - n$, and $c^{(2)}(|x - x'|; n]$ is the direct correlation function of a uniform fluid with density $n = n(x)$. The Fourier transform of the second integral in (52) gives

$$\frac{1}{2\beta} \int \int d^3x d^3x' \left[\frac{\delta(x - x')}{n(x)} - c^{(2)}(|x - x'|; n]\right]_{n_{eq}(x)} \tilde{n}(x)\tilde{n}(x') \tag{53}$$

$$= \frac{-1}{2\beta V} \sum_q \left(c^{(2)}(q; n] - \frac{1}{n(q)}\right) \tilde{n}(q)\tilde{n}(-q),$$

and comparing with (48) yields

$$f_0''(n) = \frac{-1}{\beta}\left(a(n) - \frac{1}{n}\right), \qquad f_2(n) = \frac{-b(n)}{2\beta}, \tag{54}$$

where the functions $a(n)$ and $b(n)$ are defined as coefficients of the Fourier transform of the Ornstein–Zernike direct correlation function by $c^2(q; n] = a(n(q)) + b(n(q))q^2 + \dots$. $b(n)$ is evaluated with linear response theory to find that

$$f_2(n(x)) = \frac{1}{12\beta} \int d^3x' |x - x'|^2 c^{(2)}(|x - x'|; n].$$

(55)

We can substitute Equations (55) and (50) into (49) and use the equilibrium identity $\nabla\mu = 0$ to find the integro-differential equation

$$\nabla\left[v(x) + \mu(n(x)) - f_2'(n(x))|\nabla n(x)|^2 - 2f_2(n(x))\nabla^2 n(x)\right]_{n(x)=n_{eq}(x)} = 0,$$

(56)

which determines the equilibrium density $\hat{n}_{eq}(x)$ in the presence of external potential $v(x)$ given the Ornstein–Zernike direct correlation function of uniform fluid $c^{(2)}[n(x), |x - x'|]$.

6. Conclusions

We have shown that the variational principle of classical DFT is a special case of applying the method of maximum entropy to construct optimal approximations in terms of the variables that capture the relevant physical information, namely, the particle density $n(x)$. It is worth emphasizing once again: In this paper, we have pursued the purely conceptual goal of finding how the DFT fits within the MaxEnt approach to statistical mechanics. The advantage of achieving such an insight is the potential for future applications that lie outside the reach of the current versions of DFT. As an illustration, we have discussed the already well-known example of a slowly varying inhomogeneous fluid. Future research can be pursued in three different directions: **(i)** To show that the method of maximum entropy can also be used to derive the quantum version of DFT, **(ii)** to approach the Dynamic DFT [21], generalizing the idea to non-equilibrium systems by following the theory of maximum caliber [22], and **(iii)** to revisit the objective of Section 5 and construct weighted DFTs [23,24] by using the method of maximum entropy.

References

1. Kohn, W.; Hohenberg, P. Inhomogeneous electron gas. *Phys. Rev.* **1964**, *136*, B864.
2. Mermin, D. Thermal properties of inhomogeneous electron gas. *Phys. Rev.* **1965**, *137*, A1441. [CrossRef]
3. Ebner, C.; Saam, W.F.; Stroud, D. Density-functional theory of simple classical fluids. I. surfaces. *Phys. Rev. A* **1976**, *14*, 2264. [CrossRef]
4. Evans, R. The nature of the liquid-vapor interface and other topic in statistical mechanics of non uniform, classical fluids. *Adv. Phys.* **1979**, *28-2*, 143–200. [CrossRef]
5. Kohn, W.; Sham, L.J. Self-consistent equations including exchange and correlation effects. *Phys. Rev.* **1965**, *140*, A1133. [CrossRef]
6. Zangwill, A. A half century of density functional theory. *Phys. Today* **2015**, *68*, 7. [CrossRef]
7. Jones, R.O. Density functional theory: Its origins, rise to prominence, and future. *Rev. Mod. Phys.* **2015**, *87*, 897. [CrossRef]
8. Evans, R.; Oettel, M.; Roth, R.; Kahl, G. New developments in classical density functional theory. *J. Phys. Condens. Matter* **2016**, *28*, 240401. [CrossRef] [PubMed]
9. Rowlinson, J.S. Translation of J. D. van der Waals' "The thermodynamik theory of capillarity under the hypothesis of a continuous variation of density". *J. Stat. Phys.* **1979**, *20*, 197. [CrossRef]
10. Jaynes, E.T. Information theory and statistical mechanics. *Phys. Rev.* **1957**, *106*, 620. [CrossRef]
11. Jaynes, E.T. Information theory and statistical mechanics II. *Phys. Rev.* **1957**, *108*, 171. [CrossRef]
12. Jaynes, E.T. Where do we stand on maximum entropy? In *The Maximum Entropy Principle*; Levine, R.D., Tribus, M., Eds.; MIT Press: Cambridge, MA, USA, 1979.
13. Jaynes, E.T. Predictive statistical mechanics. In *Frontiers of Nonequilibrium Statistical Physics*; Moore, G.T., Scully, M.O., Eds.; Plenum Press: New York, NY, USA, 1986.
14. Caticha, A. Entropic Physics, Probability, Entropy, and The Foundations of Physics. Available online: https://www.albany.edu/physics/faculty/ariel-caticha (accessed on 12 April 2021).
15. Rajeev, S.G. A Hamilton-Jacobi formalism for thermodynamics. *Ann. Phys.* **2008**, *323-9*, 2265–2285. Available online: https://arxiv.org/pdf/0711.4319.pdf (accessed on May 2021).

16. Balian, R.; Valentin, P. Hamiltonian structure of thermodynamics with gauge. *Eur. Phys. J. B* **2001**, *21*, 269–282. Available online: https://arxiv.org/pdf/cond-mat/0007292.pdf (accessed on May 2021).
17. Caticha, A.; Tseng, C. Using relative entropy to find optimal approximations: An application to simple fluids. *Physica A* **2008**, *387*, 6759. Available online: http://arxiv.org/abs/0808.4160v1 (accessed on May 2021).
18. Toxvaerd, S. Perturbation theory for nonuniform fluids: Surface tension. *J. Chem. Phys.* **1971**, *55*, 3116. [CrossRef]
19. Singh, Y.; Abraham, F.F. Structure and thermodynamics of the liquid—Vapor interface. *J. Chem. Phys.* **1977**, *67*, 537. [CrossRef]
20. Hansen, J.-P.; McDonald, I.R. *Theory of Simple Liquids*; Academic Press: Cambridge, MA, USA, 1976.
21. Marconi, U.M.B.; Tarazona, P. Dynamic density functional theory of fluids. *J. Chem. Phys.* **1999**, *110*, 8032. [CrossRef]
22. Jaynes, E.T. The minimum entropy production principle. *Annu. Rev. Phys. Chem.* **1980**, *31*, 579–601. [CrossRef]
23. Tarazona, P. Free-energy density functional for hard spheres. *Phys. Rev. A* **1985**, *31*, 2672. [CrossRef] [PubMed]
24. Rosenfeld, Y. Free-energy model for the inhomogeneous hard-sphere fluid mixture and density-functional theory of freezing. *Phys. Rev. Lett.* **1989**, *63*, 980. [CrossRef] [PubMed]

MDPI

St. Alban-Anlage 66

4052 Basel

Switzerland

Tel. +41 61 683 77 34

Fax +41 61 302 89 18

www.mdpi.com

Physical Sciences Forum Editorial Office

E-mail: psf@mdpi.com

www.mdpi.com/journal/psf